Contents

Teaching Science Process Skills

Jill Bailer, M.Ed.

Joyce E. Ramig, M.Ed.

John M. Ramsey, Ph.D.

Good Apple

An imprint of Paramount Supplemental Education

Executive Editor: Carolea Williams

Editor: Ema Arcellana

Copyeditor: Susan Eddy

Design: Good Neighbor Press, Inc.
        Grand Junction, CO

Inside Illustrations: Gary Mohrman

ISBN 0-86653-835-6

Printed in the United States of America

1. 9 8 7 6 5 4 3 2 1

Introduction

*"Science is built up with facts, as a house is with stones.
But a collection of facts is no more a science than a heap of
stones is a house."*

-Jules Henri Poincaré

A process skills approach to science instruction means that learning is focused
on intellectual skills rather than on content. Content, however, is not excluded
from this approach. Process skills are practiced in scientific situations which
must necessarily deal with content. Books serve as references but the emphasis
is on hands-on activities with concrete materials. Students skilled in the science
processes will be able to conduct investigations on a topic of their own choosing
with minimal teacher guidance

The activities in this resource foster understanding at the ownership level.
Students participate in the construction of their own knowledge—a concept central to current science curriculum and instruction. This constructivist concept
dictates an active curriculum in which students are provided opportunities to
collect, generate, and frame their own problems and inquiries. "Teacher talk" is
no longer the central component of science instruction. It limits student involve-
ment and results in the superficial dissemination of information. Information without understanding is inert and often forgotten. Students whose experiences are
limited solely to passive collection of information are typically unable to think
reflectively and often fail to make cognitive connections with prior knowledge.

Here are some strategies that will help the learner connect new experience to
previous knowledge. These techniques are restructuring strategies—a means of
packing and repacking ideas and explanations about a phenomenon—that can
be used throughout the curriculum.

- Have students work in small lab groups for most of the activities—they
  become more involved. When they are ready for independent investigations, students may work alone or in pairs.

- Encourage students to keep portfolios with journal entries and best work
  throughout the process skill activities. Journal forms are provided on pages
  191-192. Most classes will take from nine to twelve weeks to complete all
  the activities in the book.

- Carefully read through the entire student activity before introducing it to your class. Actively engage students using some open-ended questions or by having them predict what they think will happen within their groups. Once groups have developed consensus, predictions can be shared with the class.

- Encourage groups to come to consensus about the results of their investigations. You may wish to have them make presentations to the class ("present and defend"). The presentations may include visuals. Bring the class to consensus on the results of the activity.

- Invite students to discuss inferences in their groups by asking, "Why did this happen?" Provide each small group with a list of incomplete inferences constructed from consensus observations developed in the previous step. Invite groups to complete these inferences and present them to the class via visual aids.

- Assign one or more inferences to each small group to present and defend to the whole class.

- Bring the entire class to consensus.

Through careful monitoring of student progress and organized debriefing after each activity, you will help your students successfully internalize science process skills. Rather than lingering at the knowledge level of Bloom's Taxonomy, they will be adept at analysis, application, and evaluation of scientific knowledge and eager to undertake their own independent investigations.

Observing

Objectives

Based on experiences with observing and additional discussion, students will be expected to:

▼ Explain the role and importance of observation in the empirical nature of science.

▼ Define or select a definition for the term *observation*.

▼ Explain how observations are made.

▼ Compare and contrast the two basic modes of observation.

▼ Successfully demonstrate the skill of observing a given event and/or object.

▼ Successfully demonstrate the skill of measurement with the metric system.

▼ Successfully demonstrate the ability to use metric prefixes and units.

▼ List and defend the guidelines necessary for making good observations.

Observing

All human beings share a desire to explore and understand the world around us. Science developed out of this curiosity. Science is based on empiricism—a search for knowledge based on experimentation and observation. Sometimes observations are made directly with the human senses. Sometimes technological devices, such as microscopes, are used to help. Scientists believe that without meaningful observations based on thoughtful use of experience, the evidence or data needed to understand a problem will be incomplete.

There are two basic types of observations—*qualitative* and *quantitative*. Qualitative observations describe and quantitative observations measure. Scientists usually try to use quantitative observations because they are more precise. Thoughtful observation communicates. It is clear and detailed and it takes practice. The following activities will give you a chance to practice making clear, thoughtful observations.

Teaching Science Process Skills © 1995 Good Apple

Confection Connection

Materials (per student)

- one piece of wrapped candy
- introductory information on observing on page 8
- worksheet on page 10

Active Engagement Before the Task

Distribute the introductory information on the skill of observing to each student. Individually, in small groups, or as a class read and discuss the material. Be sure students understand the idea of empiricism. Actively engage students' minds in preparation for the process skill activity by asking the following questions.

- Which of your five senses would you least like to live without? Why?

- Does your sense of smell enhance the experience of eating? Explain your answer.

- Do you think you would enjoy your favorite food as much if you were unable to see it? Why or why not?

Process Skill Activity

Students will work individually on this activity. Have students close their eyes and place an open hand, palm up, on their desks. Place a piece of wrapped candy in each student's hand. Remind students to keep their eyes closed. The students should focus only on the sense of touch. When exploration is completed, direct students to open their eyes. Invite students to explore independently and record their observations for the other senses. They may wish to share their descriptive paragraphs with the class.

In this activity students will make qualitative observations of a piece of candy using their five senses. They will then write a descriptive paragraph of the item they have observed without mentioning its name.

Name _____   Date _____

Confection Connection

This activity gives you the opportunity to make some observations. Remember that observations are based on the five senses and should be factual.

 PART ONE Close your eyes while your teacher distributes an object to observe. Use only your sense of touch at this time. Use your other senses only when your teacher directs. Record your observations below.

Sense of Touch: Describe what you feel.

_____

Sense of Hearing: Listen as you unwrap the object. Describe what you hear.

_____

Sense of Smell: Describe what you smell.

_____

Sense of Sight: Describe what you see.

_____

Sense of Taste: Place the object in your mouth. Describe what you taste.

_____

PART TWO Write a detailed paragraph describing the object to a young child. Do not mention the name of what you are describing.

_____

_____

_____

_____

_____

_____

_____

_____

Teaching Science Process Skills © 1995 Good Apple

Penny Observations—A Close-Up Look

Teacher Notes

Materials (per student)

- one penny
- magnifying lens
- vinegar
- salt
- paper cup
- worksheet on pages 12-13

Active Engagement Before the Task

Actively engage students' minds in preparation for the process skill activity by asking the following questions.

- There are six things that appear on every single United States coin. Make a list of what you think they might be. (date, denomination, and the words *United States of America, In God We Trust, Liberty* and *E Pluribus Unum.*)

- What metal are pennies made of? What do you think causes the metal to tarnish?

- What do you think might happen when a penny is placed in a vinegar/salt/water solution? Would the same thing happen to a dime or a nickel?

Process Skill Activity

Do not distribute pennies until Steps 1 and 2 are complete. Set out materials for the experiment for easy access when students are ready to use them. This activity is more appropriately done by individual students. Once students have decided on an investigation question (Part Four), have them bring their investigation plan to you for approval. Results can be shared with the class.

In this activity, students will make three types of observations of a penny—from memory, from short observation and from long observation. They will perform a simple experiment using a penny and design a simple investigation to find out something they do not yet know about pennies.

Name _____ Date _____

Penny Observations—A Close-Up Look

Your task is to make three different types of observations about a penny.

 PART ONE First, you will make observations about a penny from memory. Next, you will observe a penny for one minute and then record your observations. Finally, you will make continuous, direct observations.

Step 1

From memory, draw the front and back of a penny in the circles. Do not look at a penny. Tell about what you drew.

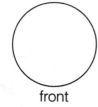 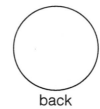

front   back

_____

_____

Step 2

Observe a penny for one minute and then place it out of sight. Draw the front and back of the penny in the circles. Compare your drawings to those you drew in Step 1. Record any similarities or differences.

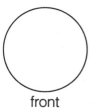 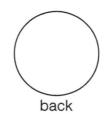

front   back

_____

_____

Step 3

Observe a penny for as long as you need. Use a hand lens if you wish. Draw the front and back of the penny in the circles. Tell how your drawings changed over the three steps.

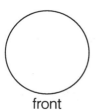 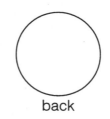

front   back

_____

_____

Teaching Science Process Skills © 1995 Good Apple

 You may have observed that your penny is tarnished. Place a small amount of vinegar in a paper cup and add a pinch of salt. Now add your penny. Remove the penny after several minutes and observe it carefully. In the space below, describe what happened. Include in your description what the penny looked like both before and after placing it in the vinegar/salt solution.

_____

_____

 What would you like to find out about pennies? Write your questions in the space below.

_____

_____

 Design an investigation to answer one of your questions for Part Three. Write the question, your plan, and the results in the spaces below.

Question _____

_____

Step-By-Step Investigation Plan _____

_____

_____

Investigation Results_____

_____

_____

The Metric System

Materials (per student)

- introductory information on quantitative observations on page 15
- table of basic metric prefixes on page 16
- worksheet on page 17

Active Engagement Before the Task

Distribute the introductory information on quantitative observations and the table of basic metric prefixes and quantities to each student. Individually, in small groups, or as a class read and discuss the material. An overhead transparency of the table of basic metric prefixes may be useful. Actively engage students' minds in preparation for the process skill activity by asking the following questions.

- *Centum* is Latin for one hundred. How can you relate this to your chart of basic metric prefixes? What other English words can you think of that derive from the Latin word centum? (century, centennial, centipede)

- Which system of measurement seems the most "user-friendly" — customary U.S. measurement or the metric system? Why?

In this activity, students will learn the difference between qualitative and quantitative observations. In addition, they will become familiar with the metric system—its derivation, rationale, and conversion techniques.

Process Skill Activity

Students will work individually on this activity as they complete the exercises in metric conversion.

Answer Key for Practice With Metric Conversions

6200 grams	0.0672 kilometers
78,360 milliliters	Distance: meter
1230 centimeters	From home to school: kilometers
0.0032 kilometers	Length of the classroom: meters
7.63 centimeters	Width of this page: centimeters
143,200 millimeters	
0.312 grams	Volume: liter
0.0153 liters	Volume of a large jug: liters
7500 milligrams	Liquid medicine: milliliters
6210 centimeters	
53,500 milliliters	Mass: gram
12,300 meters	Your mass: kilograms
0.0794 liters	Mass of a pin: milligrams
0.0000439 kilograms	

Quantitative Observations

▼ When you observed the candy and the penny, you were making qualitative observations that described observable characteristics. The candy was sweet and the penny was brown. However, in order to think like a scientist, it is necessary to make some quantitative observations—observations that require measurement or numerical calculation. An example would be that the Lincoln Memorial on a penny has twelve columns. Make as many quantitative observations for a penny as you can. Record your observations.

_____

_____

_____

_____

_____

_____

_____

_____

_____

Table of Basic Metric Prefixes and Quantities

A quantitative measurement system has been developed that is used in most of the world. It is called the International System or, more commonly, the metric system. The metric system was designed to relate mass, distance, and volume for one substance—pure water. This is how it works. Imagine a small box that is exactly one centimeter long, one centimeter wide, and one centimeter high. Its volume is one cubic centimeter (cc). If water is added to this container until it is full, that amount of water would be one milliliter and have a mass of one gram. This assumes that water is at a standard (normal) temperature and a standard (normal) air pressure.

The metric system is based on multiples of ten. This makes it very easy to change from one unit to another and it makes it easier to use very large or small numbers.

The basic units of the metric system are the liter, a measure of volume; the meter, a measure of distance; the gram, a measure of mass; and degrees Celsius, a measure of temperature.

Prefix	Quantity	Symbol	Example
kilo	1000	k	1 kilogram = 1000 grams 1 kg = 1000 g
hecto	100	h	1 hectogram = 100 grams 1 hg =100 g
deca	10	dk	1 decagram = 10 grams 1 dkg = 10 g
base unit	1	g, l, m,	1 gram = 1 g 1 liter = 1L 1 meter = 1 m
deci	0.1 (1/10)	d	1 decigram = 0.1 grams 1 dg = 0.1 grams
centi	0.01 (1/100)	c	1 centigram = 0.01 grams 1cg = 0.01 g
milli	0.001 (1/1000)	m	1 milligram = 0.001 grams 1 mg = 0.001 g

Name _____ Date _____

The Metric System

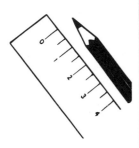

Change the following numbers to the appropriate units of measurements.

6.2 kilograms = _____ grams

78.36 liters = _____ milliliters

12.3 meters = _____ centimeters

3.2 meters = _____ kilometers

76.3 millimeters = _____ centimeters

143.2 meters = _____ millimeters

312 milligrams = _____ grams

15.3 milliliters = _____ liters

7.5 grams = _____ milligrams

62.1 meters = _____ centimeters

53.5 liters = _____ milliliters

12.3 kilometers = _____ meters

79.4 milliliters = _____ liters

43.9 milligrams = _____ kilograms

67.2 meters = _____ kilometers

Write the names of the basic metric units used to measure each of the following numbered items. Then, next to each lettered item, tell which subunit of these measurements would be most practical to use.

Distance _____

a. From home to school _____

b. Length of the classroom _____

c. Width of this page _____

Volume _____

a. Volume of a large jug _____

b. Liquid medicine _____

Mass _____

a. Your mass _____

b. Mass of a pin _____

Measurement of Length

Materials (per student)

- metric ruler
- meter stick
- notebook paper
- tape
- book
- scissors
- worksheet on page 19

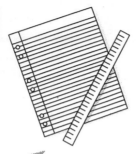

In this activity, students will practice making careful, quantitative observations of length.

Active Engagement Before the Task

Actively engage students' minds in preparation for the process skill activity by asking the following questions.

- What situations or professions can you think of in which it is important to be extremely accurate in measurements of distance?

- Explain the wisdom of the carpenter's motto, "Measure twice, cut once."

Process Skill Activity

Have students work in groups to complete the worksheet; however, encourage students to come up with answers individually before comparing them with the rest of the group.

Name _____ Date _____

Measurement of Length

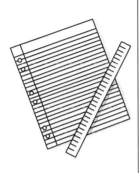

In this and the following three activities, you will have the opportunity to practice making careful, quantitative observations. You will be using specific equipment to make metric measurements of length, mass, volume, and temperature. Make sure your work is accurate and be sure to label each measurement with its correct unit. Metric length is measured in meters, centimeters, and millimeters with a meter stick or metric ruler.

1. Measure the length of the line below in cm and mm.

_____ _____ cm or _____ mm

2. Cut a piece of notebook paper that measures 50 mm wide and 250 mm long. Then cut a piece of notebook paper that is 25 cm long and 5 cm wide. Hint: Find a way to save work here.

3. Find the area (A = L x W) of the rectangle constructed in Step 2. _____

4. Cut a square of notebook paper that is 4.5 cm on each side.

5. Measure the length, width, and thickness of a book. Choose the best unit.

length = _____ width = _____ height = _____

6. Measure the length, width, and height of a desk or table in the room. Choose the best unit.

length = _____ width = _____ height = _____

7. Measure the perimeter of this sheet of paper (P = 2L +2W).

Choose an appropriate unit. _____

8. Draw a 98 mm line in the space below.

9. Measure the length of your pen or pencil. Choose the appropriate unit. _____

Teaching Science Process Skills © 1995 Good Apple

$\mathcal{M}$easurement of Volume

Materials (per group)

- various graduated cylinders
- four graduated cylinders with red, blue, green, and yellow water
- beaker of water
- one liter container
- measuring cup
- scissors
- tape
- container (can or bottle)
- box pattern sheet on page 21
- worksheet on page 22

▼ *In this activity, students will practice making careful, quantitative measurements of volume.*

Active Engagement Before the Task

Actively engage students' minds in preparation for the process skill activity by asking the following questions.

- Can you name a situation or profession in which it is important to be extremely accurate in measurements of volume?

- Do you think glass and plastic containers would have different effects on the liquids they contain?
  Why or why not?

- Can you think of any relationship between the U.S. customary system of volume measurement and the U.S. customary system of linear measurement?

Process Skill Activity

Have students work in groups for this activity. Duplicate the box pattern to provide one small box for each student. Review the protocol for measuring with a graduated cylinder. Have groups present their answers individually when all have finished. Note that you will need to know the volume of water in the graduated cylinders beforehand, particularly if you choose to make amounts different for each group.

Box Patterns

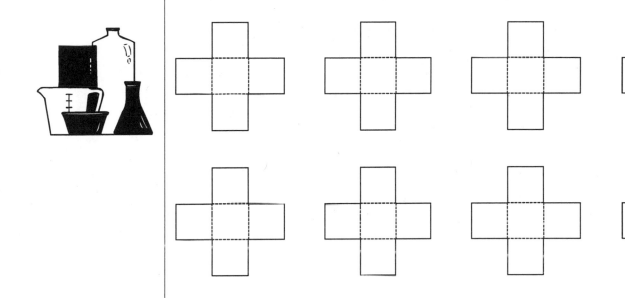

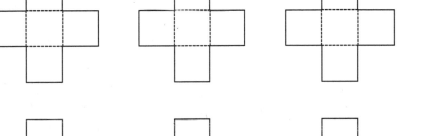

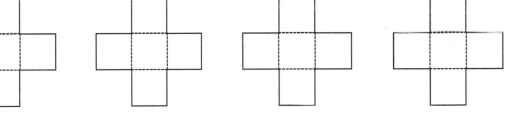

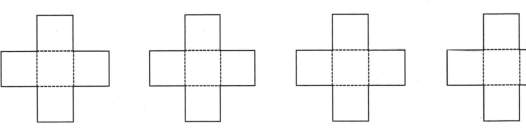

Measurement of Volume

Procedure or Protocol for Measuring with a Graduated Cylinder

▶ Place the graduated cylinder on a flat, solid surface.

▶ Your eyes must be at the level of the graduated cylinder. Do not pick it up.

▶ Pour the liquid to be measured into the graduated cylinder.

▶ The proper volume is the line that is touched by the bottom of the curve (or meniscus) of the liquid in a glass cylinder. In a plastic graduated cylinder, there will be no meniscus.

	Volume
1. Record the volume of the graduated cylinder with red water.	_____
2. Record the volume of the graduated cylinder with blue water.	_____
3. Record the volume of the graduated cylinder with green water.	_____
4. Record the volume of the graduated cylinder with yellow water.	_____
5. Pour exactly 95 ml of water into a graduated cylinder.	
6. Pour exactly 52 ml of water into a graduated cylinder.	
7. How many ml in 1 L?	_____
8. How many ml in 1 cup?	_____
9. Find the volume of a container.	_____
10. Using the pattern provided, construct a cubic centimeter. Use tape to hold the shape.	

$\mathcal{M}$easurement of Mass

Materials (per group)

- balance
- any three available objects
- small graduated cylinder
- plastic bags
- sugar
- dissolving antacid tablet
- 50 ml water
- paper towels
- container of water
- worksheet on page 24

Active Engagement Before the Task

Actively engage students' minds in preparation for the process skill activity by asking the following questions.

- What situations or professions can you think of in which it is important to be extremely accurate in measurements of mass?

- Does air have weight? Do different types of air have different weights? How could you find out?

- Do you think the temperature of an object influences its weight? Why or why not?

Process Skill Activity

Review the protocol for measuring with a balance scale with your students. If this is a new skill, demonstrate the procedure. Have students work in groups to complete the worksheet on measurement of mass. Have students fill in the three blanks at the beginning of the table with the name of three objects they would like to calculate the mass of. The group should reach consensus before recording their answers. If you choose, they can present their findings to the class.

In this activity, students will practice making careful, quantitative observations of mass.

Measurement of Mass

Procedure or Protocol for Measuring with a Balance Scale

▼ Place the balance on a flat, solid surface.

▼ Place the object to be measured on the left pan. Note: If chemicals or liquids are to be measured, protect the balance surface by using a container to hold the chemicals. Place the container on the left pan.

▼ Zero the balance.

▼ Place standard (known) masses on the right pan until the balance arms are horizontal (the pans are at the same height).

▼ Total the masses required to balance the unknown mass.

	Object	Mass
1.	A _____	_____
2.	B _____	_____
3.	C _____	_____
4.	Small graduated cylinder	_____
5.	10 ml of water	_____
6.	Plastic bag containing 50 g of sugar	_____
7.	The difference in mass between a dry and a moist paper towel	_____
8.	Mass of 50 ml of hot water and 50 ml of cold water	_____ _____
9.	Mass of an antacid tablet before and after dissolving in 50 ml of water	_____ _____

$\mathcal{M}$easurement of Temperature

Teacher Notes

Materials (per group)

- thermometers
- room temperature tap water
- hot water
- ice water
- ice/salt water mixture
- 100 ml water in screw-top jar
- modeling clay
- refrigerator oooooo
- worksheet on page 26

Active Engagement Before the Task

Actively engage students' minds in preparation for the process skill activity by asking the following questions.

- What situations or professions can you think of in which it is important to be extremely accurate in measurement of temperature?

- What do you think happens to modeling clay that has been held in your hand? What causes these changes?

- Can there be more than one air temperature inside one room? Why or why not?

Process Skill Activity

Review the protocol for measuring with a thermometer with your students. Have them work in groups to complete the worksheet. Groups may wish to divide up the tasks among the members. Findings may be presented to the class.

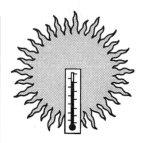

In this activity, students will practice making careful, quantitative observations of temperature.

Name _____ Date _____

Measurement of Temperature

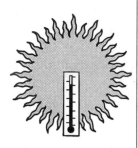

Procedure or Protocol for Measuring with a Thermometer

▶ Place the thermometer in the substance to be measured for one minute or as designated.

▶ Do not remove the thermometer until you have read the temperature.

▶ Use the Celsius scale.

		Temperature
1. room temperature tap water		_____
2. hot water		_____
3. room air		_____
4. inside of a refrigerator		_____
5. window air		_____
6. ice water		_____
7. ice/salt water mixture		_____
8. inside of your fist		_____
9. 100 ml of water before and after vigorous shaking	_____	_____
10. modeling clay before and after 25 squeezes	_____	_____

Teaching Science Process Skills © 1995 Good Apple

Mixing Water Activity

Materials (per group)

- thermometer
- cold water
- hot water
- warm water
- graduated cylinders
- container for mixing
- worksheet on page 28

Active Engagement Before the Task

Actively engage students' minds in preparation for the process skill activity by asking the following questions.

- What do you think would happen if equal parts of 10°C water and 60°C water were mixed?

- What would make a pool containing 22°C water feel cool on one day and warm on another?

- What do you think causes your temperature to rise when you are sick? Can you lower your temperature by having a cold drink? Why or why not?

Process Skill Activity

Have students work in groups to complete the mixing water activity. Remind them that they are making quantitative observations and that their work should be as accurate as possible. Groups should come to consensus concerning the reason why results were or were not as expected. Have groups share their findings and then bring the class to consensus.

In this activity, students will make quantitative observations after mixing various amounts of water of different temperatures.

Name _____ Date _____

Mixing Water Activity

1. Measure 40 ml of cool water and 40 ml of warm water in separate graduated cylinders.

2. Record the temperatures of each in the table below.

3. Mix the cool and warm water together. Record the temperature of the mixture.

4. Repeat the steps above for the remaining combinations as specified in the table.

Amount of Water	Temperature of Cool Water	Temperature of Warm Water	Temperature of Mixed Water
40 ml of cool water and 40 ml of warm water			
30 ml of cool water and 50 ml of warm water			
20 ml of cool water and 50 ml of warm water			
10 ml of cool water and 70 ml of warm water			

Were your results what you expected? Why or why not?_____

_____

_____

_____

Teaching Science Process Skills © 1995 Good Apple

Airplane Activity

Materials (per student)

- balance
- ruler
- paper
- area for plane flying
- worksheet on page 30

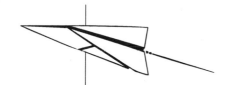

Active Engagement Before the Task

Actively engage students' minds in preparation for the process skill activity by holding up a paper airplane you have made and asking the following questions.

- What are some qualitative observations you could make about this plane?

- What changes could I make that might increase the flying distance of this plane?

- What do you think is the single most important element in a successful paper plane?

Process Skill Activity

Students will work individually on this activity. Remind them that they are making quantitative observations and that their work should be as accurate as possible. Once it has been determined which plane flies the farthest, you may want to analyze with the class why this is so. Students may want to modify their planes based on the discussion.

In this activity, students will make quantitative observations and illustrations of paper airplanes they create.

Name ______________________________ Date ______________

Airplane Activity

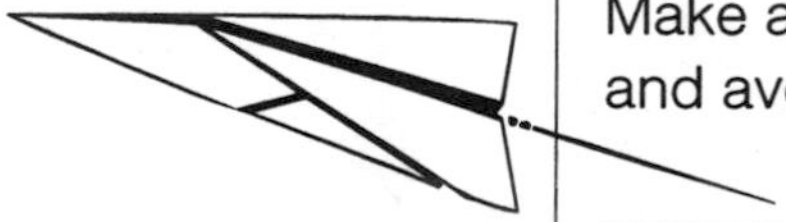

Make a paper airplane. Record at least six measurements including mass, length, and average flying distance. Remember to include units of measurement.

Mass	______________
Length	______________
Average flying distance	______________
______________	______________
______________	______________
______________	______________
______________	______________
______________	______________
______________	______________

Draw your airplane.

Candle Activity

Teacher Notes

Materials (per group)

- birthday candle
- oil-base clay (to hold candles)
- matches (used under close adult supervision)
- ruler
- balance

Active Engagement Before the Task

Actively engage students' minds in preparation for the process skill activity by asking the following questions.

- Do you think the flame on a candle has weight? Why or why not? How could you find out?

- What do you think happens to the wax on dripless candles?

- If wax drips down the side of a burning candle, do you think the weight of the candle will change after one minute? If so, what would the change be?

Process Skill Activity

Have your students work in groups on this activity. Remind them to be as accurate as possible when making quantitative observations and to use caution when lighting and disposing of matches. The groups may wish to subdivide between qualitative and quantitative observations. Have groups share their observations with the class. This would be a terrific activity to do on a student's birthday. Substitute cupcakes for the clay.

In this activity, students will make qualitative and quantitative observations of a birthday candle before, during, and after burning.

Name _____ Date _____

Candle Activity

Make qualitative and quantitative measurements of a small candle both before and after it has burned for two minutes. Use caution when lighting matches and dispose of them properly. Anchor the candle in a ball of modeling clay.

Qualitative Observations

Before burning _____

_____

During burning _____

_____

After burning _____

_____

Quantitative Observations

Observations	Before Burning	After Burning

Inferring

Objectives

Based on experiences with inferring and additional discussion, students will be expected to:

▼ Define or select a definition for the skill of inferring.

▼ Describe the difference between an observation and an inference.

▼ Effectively communicate the extent of risk involved in making inferences.

Inferring

▼ The world is a lot more enjoyable when the things that happen around us are understandable. When we observe patterns that are similar or events that remind us of experiences we've had previously, a new experience can be appreciated more fully. The explanations we use to depict events we experience are often called *inferences*. Inferences are based on observations and are simply explanations of observations. The ability to infer helps us make sense of our environment.

The only rule of inferring is to be logical. Inferences are always tentative. They are not final explanations of an observation. Sometimes there are several logical inferences for a given observation and you cannot be sure which inference best explains the observation. In such instances, you will need additional information. Inferences are often changed when new observations are made.

Teaching Science Process Skills © 1995 Good Apple

Making Inferences

Materials (per student)

- introductory information on inferring on page 34
- worksheet on pages 36-38

Active Engagement Before the Task

Distribute the introductory information on the skill of inferring to each student. Individually, in small groups, or as a class read and discuss the material. Discuss the importance of making logical inferences. Actively engage students' minds in preparation for the process skill activity by doing a few observations and inferences as a class.

- You observe a dejected-looking student leaving the principal's office. What is your inference?

- You see several people leaving a movie theater red-eyed and blowing their noses. What is your inference?

- Show students an appropriate photograph or illustration from a magazine or distribute the worksheets and have students look at the first illustration. Ask students for their inferences.

Process Skill Activity

Students will complete the written observations/inferences first. Then, invite students to make inferences based upon the pictures on pages 37-38. Have groups come to consensus and share them with the class.

In this activity, students will practice making inferences both from written statements and from pictures.

Name _____ Date _____

$\mathcal{M}$aking Inferences

Read the following observations. Then make inferences that explain each observation. Remember, there may be more than one logical explanation.

Observation 1: You observe that the sky at noon is darkening.

Your inference: _____

Observation 2: The principal interrupts class and calls a student from the room.

Your inference: _____

Observation 3: All middle school students are bringing lunch from home.

Your inference: _____

Observation 4: A former rock-and-roll band member has poor hearing.

Your inference: _____

Observation 5: You leave a movie theater and see that the street is wet.

Your inference: _____

Observation 6: During a handshake, you feel that the palm of the individual's hand is rough and hard.

Your inference: _____

Observation 7: The classroom lights are off.

Your inference: _____

Observation 8: A siren is heard going past the school.

Your inference: _____

Teaching Science Process Skills © 1995 Good Apple

Name _____ Date _____

Observations and Inferences

Use the pictures that follow to make observations and inferences.

Observations: _____

_____

_____

Inferences: _____

_____

_____

_____

Observations: _____

_____

_____

Inferences: _____

_____

_____

_____

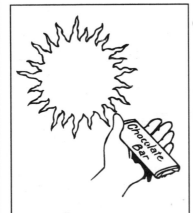

Observations: _____

_____

_____

Inferences: _____

_____

_____

_____

Name _____ Date _____

Observations: _____

_____

_____

Inferences: _____

_____

_____

_____

Observations: _____

_____

_____

Inferences: _____

_____

_____

_____

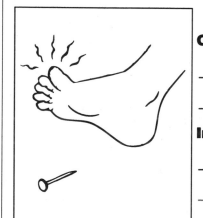

Observations: _____

_____

_____

Inferences: _____

_____

_____

_____

$\boldsymbol{J}$t's on the Paper

Materials (per group)

- goldenrod paper cut into quarters or eighths
- colored paper other than goldenrod cut into quarters or eighths
- baking soda solution (2-3 tsp. baking soda in 1 pint water)
- spray bottle filled with vinegar
- marker
- cotton swab or paint brush
- worksheet on pages 41-43

In this activity, students will observe an acid-base reaction and make inferences based on their observations.

Some kinds of goldenrod paper are made with a dye that is yellow in acids and a deep pink-red in bases. One brand of paper that works quite well is Astro Bright "Galaxy Gold." "Goldenrod" by Mead Paper Company also works well. Use your finger, a cotton swab, or a paint brush moistened with the baking soda solution to write something, such as an X or a question mark, on one small piece of goldenrod paper for each group. The paper should turn a pink-red color where you apply the solution. When sprayed with vinegar by the students, the color should revert to yellow. Use a magic marker to write the same X or question mark on a different color of paper.

Active Engagement Before the Task

Review with your students how an inference is made and remind them that there is not a right or a wrong inference. Actively engage students' minds in preparation for the process skill activity by asking the following questions.

- What might cause "disappearing ink" to disappear or reappear?

- What ingredients could be in a recipe for disappearing ink?

- Think back to what vinegar did to a penny. What do you think it might do to colored paper?

Process Skill Activity

Students will work in groups on this activity. Remember to encourage groups to come to consensus before sharing their observations and inferences with the class.

Part Four Answer Key

1. dyes
2. pink-red
3. yellow or goldenrod

It's on the Paper

You are going to practice making more inferences about things you observe. Keep in mind that an inference must make sense and provide an explanation for your observation.

PART ONE You have been given two pieces of paper with some writing on them. Record your observations about these papers.

_____

_____

_____

Make some inferences to explain your observations. _____

_____

_____

_____

 PART TWO Using a spray bottle provided by your teacher, spray some of the liquid on both papers. Record what you observe.

_____

_____

_____

What can you infer from your observations? _____

_____

_____

PART THREE Make more inferences about what you observed by completing the following statements.

There is writing on the paper because _____

_____

The papers are colored because_____

_____

The writing is colored because _____

_____

The writing disappeared because _____

_____

The writing did not disappear because _____

_____

PART FOUR The story below explains what happened with the writing on the papers. Certain key words, however, have been omitted. Write the missing word(s) that will make each sentence correct.

About the Paper

Paper is colored with special chemicals called _____. Some
kinds of goldenrod paper are made with a dye that is yellow in acids and
pink-red in bases. When a base, such as baking soda in solution with water,
is put on this kind of goldenrod paper, the paper turns a

_____ color. When an acid, such as vinegar, is sprayed on
the pink-red color, the paper is changed back to _____.
Materials that can be used to test whether a substance is an acid or base
are called indicators.

Teaching Science Process Skills © 1995 Good Apple

PART FIVE

In your own words, explain the color change on the paper. Include drawings if you think they will help explain your ideas.

_____

_____

_____

_____

_____

_____

_____

_____

_____

*I*t's in the Liquid

Materials (per group)

- water colored blue
- rubbing alcohol colored red
- toothpicks
- 2 medicine droppers
- plastic bag
- worksheet on pages 46-48

Active Engagement Before the Task

Actively engage students' minds in preparation for the process skill activity by asking the following questions.

- Why do you think a belly flop is so painful?

- Do you think that a belly flop into a pool filled with some liquid other than water would be less painful? Why or why not? (water has extremely high surface tension—it is lower in other liquids)

- What might the surface tension of water have to do with the formation of bubbles?

In this activity, students will practice making inferences by observing the effect of plastic and rubbing alcohol on the surface tension of water.

Process Skill Activity

Have students complete Part One through Part Four in their groups. Once all groups have completed these exercises, have them come to consensus and share their inferences with the class. Part Five provides an explanation for what has occurred and should be distributed after group consensus has been shared. You may choose to have students complete Parts Five and Six on their own.

Part Five Answer Key (answers may vary)

1. surface tension
2. stick together
3. beads up
4. toothpick
5. plastic
6. break apart
7. broke
8. not follow the toothpick

It's in the Liquid

PART ONE ▶ Place a dime-sized drop of the blue liquid on the plastic bag. Use a toothpick to move it around.
Record your observations.

_____

_____

_____

What inferences can be made based on your observations?

_____

_____

_____

 Here are more opportunities to use your skills of observing and inferring.

PART TWO ▶ Place a dime-sized drop of the red liquid on the plastic bag. Do NOT allow the red and blue liquids to mix. Use the toothpick to move the red liquid around. What did you observe?

_____

_____

_____

What inferences can be made based on your observations?

_____

_____

_____

Name _____ Date _____

_____

▶ **PART THREE** ▶ Allow the drops of blue and red liquid to mix on the plastic bag. Observe the liquids as they mix.

Record your observations.

_____

_____

What inferences can be made about what you observed?

_____

_____

_____

▶ **PART FOUR** ▶ Make more inferences about what you observed by completing the following statements.

The blue liquid stayed in a drop shape because _____

_____

The blue liquid followed the toothpick because _____

_____

The red liquid was spread out because _____

_____

The red liquid did not follow the toothpick because _____

_____

When the liquids mixed, the blue liquid spread out because _____

_____

When the liquids mixed, they fizzed because _____

_____

Name _____ Date _____

 PART FIVE ▶ The story below explains what happened when the red and blue liquids mixed. Certain key words or phrases, however, have been omitted. Fill in the missing word(s) or phrase(s) that will make each sentence correct.

About the Liquids

The blue liquid is a combination of water and blue food coloring. Water is a very special liquid. It has extremely high *surface tension*. Water molecules like to stick together and they hold on to each other the tightest at the surface. The reason it hurts when you do a belly flop into a swimming pool is because of _____.
1

Although water molecules like to _____, there are
2
some things to which they don't like to stick. One of these things is plastic. When water is placed on a plastic bag, it _____. The water follows
3
the toothpick around, because water likes to stick to the _____
4
more than to the _____.
5

When other substances, such as soap or rubbing alcohol are mixed with water, the water molecules can't stick together as well. The red liquid contained rubbing alcohol that caused the water to

_____. The water and alcohol would not follow
6
the toothpick because the alcohol _____ the surface tension.
7
When the red and blue liquids were mixed, the surface tension of the water was broken. That caused the purple drop of water and alcohol to

_____.
8

PART SIX ▶ In your own words, explain what happened with the blue and red liquids.

_____

_____

_____

_____

Teaching Science Process Skills © 1995 Good Apple

Mysterious Journeys in the Life of a Raisin

T e a c h e r   N o t e s

Materials (per group)

- 250 ml beaker
- plastic spoon
- 100 ml water
- baking soda
- 30 ml vinegar
- raisins
- graduated cylinder
- worksheet on pages 51-57

Active Engagement Before the Task

Actively engage students' minds in preparation for the process skill activity by asking the following questions.

- What do you think causes some cakes of soap to float?

- Can you think of a way to make something float that has already sunk?

- Do you think people have different "floatability" factors?

Process Skill Activity

Students will work in groups on Parts One through Four of the activity. They should measure out their own supplies using a graduated cylinder from a central source you have provided. Part Five is an optional independent investigation for which students will provide their own materials. You may find it necessary to do Parts Six and Seven during a second class period if scheduling is inflexible. You may decide to do Part Five on a day of its own, before or after you get to Parts Six and Seven. Remind students that inferences are explanations of observations and should be logical. Have groups share their findings with the class.

In this activity, students will make qualitative observations of and inferences about the various activities of raisins in a solution of water, baking soda, and vinegar.

Part Six Answer Key (answers may vary)

About Force
lifting

The Rise (about bubbles)
1. baking soda
2. rose
3. accumulated on raisins
4. rising bubbles

The Fall
1. popped
2. liquid
3. lifting

The Flip
1. bubbles
2. sink
3. popped

The End
1. heavy
2. bubbles

$\mathcal{M}$ysterious Journeys in the Life of a Raisin

 PART ONE ▶ You are going to observe the behavior of raisins in a solution and make inferences based on what you observe. Use as many of your senses as you can while making your observations. Make sure your observations are clearly written.

1. Pour 100 ml tap water into the 250 ml beaker.
2. Add 1 teaspoon of baking soda.
3. Add several raisins.
4. Add 15 ml vinegar.
5. Observe what happens for several minutes. Record your observations below.
6. Add another 15 ml vinegar.
7. Again, observe for several minutes. Write what happened below.

Observations: _____

_____

_____

_____

_____

 PART TWO ▶ Make a drawing that shows the activity of the raisins in the liquid. Be sure to label the important parts.

Teaching Science Process Skills © 1995 Good Apple

Name _____ Date _____

 PART THREE ▶ **1.** Consider the following observation.
The water smelled funny.

This observation is not written clearly. It doesn't tell much about what the observer experienced.

Consider another observation.
The solution in this activity smelled like vinegar.

This observation is written clearly. It gives a good description of what was observed.

2. Read the following observations. Circle the letter next to the one that is the most clearly written.

    A. The raisins moved around.

    B. Three raisins with bubbles attached rose to the top of the solution.

    C. There were raisins and lots of bubbles. Both the raisins and the bubbles were moving up and down.

3. Review the observations that you wrote in Part One. Find one that you believe is written clearly and write it on the lines below.

_____

_____

Find one of your observations that needs improvement. Write it on the lines below.

_____

_____

Rewrite the observation to make it more descriptive and more exact.

_____

_____

Teaching Science Process Skills © 1995 Good Apple

PART FOUR ▶ Make some inferences about your observations. Remember that an inference is a tentative explanation. Explain the activities of the raisins in the solution by completing the following statements.

Some of the raisins floated because_____

_____

After rising, most of the raisins sank because_____

_____

After rising, some bubbles attached to the raisin popped at the top of the solution

because _____

_____

After rising, some raisins turned or flipped over as they sank because_____

_____

_____

After a period of time, most raisins stopped rising because _____

_____

Some raisins did not rise because _____

_____

Some raisins bumped and turned around on the bottom of the container because

_____

_____

Bubbles are round (spherical) in shape because _____

_____

Raisins with a lot of bubbles rise but raisins with fewer bubbles don't rise because

_____

_____

 Try your own investigation. Think of other objects that might act like the raisins and other solutions with gas bubbles. Plan your investigation and bring in the materials you will need to do the experiment. Write your plan, observations, and inferences clearly in the spaces provided.

Investigation Plan

_____

_____

_____

_____

_____

Observations

_____

_____

_____

_____

_____

_____

Inferences

_____

_____

_____

_____

_____

_____

Teaching Science Process Skills © 1995 Good Apple

PART SIX ▶ The story below explains the activity of the raisins. However, certain key words or phrases have been omitted. Write the missing word(s) or phrase(s) that will make each sentence correct.

About Force

Raisins float and sink because of two kinds of forces. One kind of force is a lifting force called *buoyancy*. The second kind of force is a sinking force called *gravity*. When an object is placed in a liquid, the liquid and the object push on each other. In the case of the baking soda and vinegar solution, the heavier vinegar solution pushed the lighter bubbles upward. These rising bubbles created a lifting force. The bubbles accumulated on the raisins. As more and more bubbles accumulated on the raisins, there was a greater _____ force present.

The Rise (about bubbles)

Bubbles were made by a chemical reaction of the two main ingredients in the solution, _____ and vinegar. The bubbles consist of a gas called carbon dioxide. Since the bubbles were lighter than the vinegar solution, they _____.
₁
₂
This action of the bubbles created a buoyant force. Sometimes the bubbles bumped into raisins and sometimes they _____. If
₃
enough bubbles accumulated on a raisin, the lifting force made by the

_____ caused the raisin to rise. As a result, the raisins
₄
floated to the top of the solution.

The Fall

At the surface of the solution, some of the bubbles near the top of the raisins

_____. This happened because the bubbles were no longer surrounded
    1

by the _____. Now the floating raisins had fewer bubbles. With fewer
      2

bubbles, the raisins had less _____ force. The raisins sank to the bottom
         3

of the container because the lifting force was less than the sinking force.

The Flip

Some raisins flipped as they began to sink. This happened because

_____ on only one side of the raisins popped. This caused the lifting and
    1

sinking forces to be unbalanced. The side of the raisins that lost the bubbles began

to _____. Some raisins did not behave this way. They simply sank straight
   2

down, because the bubbles _____ equally on all sides of the raisins.
       3

Some raisins floated to the top and stayed there because there were still enough

unpopped bubbles under the raisins.

The End

Some raisins did not rise at all because they were too _____.
        1

They acquired some bubbles but they could not get enough lift to overcome the

sinking force. After a period of time the raisin activity stopped. This occurred

because the ingredients were all used up and no more _____ were
        2

chemically produced.

Teaching Science Process Skills © 1995 Good Apple

Name _____ Date _____

PART SEVEN

In your own words, explain the activities of the raisins in the baking soda and vinegar solution. Include drawings if you think they will help explain your ideas.

_____

_____

_____

_____

_____

_____

_____

_____

_____

_____

_____

_____

$\mathcal{S}$weet but Sticky Observations and Inferences

Materials (per student)

- balance
- ruler
- variety of brands and flavors of chewing gum (for independent investigations)
- worksheet on pages 60-62

Active Engagement Before the Task

Review the difference between quantitative and qualitative observations and briefly explain the task students will be undertaking. Hold up a stick of chewing gum and ask students to make three or four qualitative and quantitative observations of the gum. (One quantitative observation should be the mass.) Write observations on the board. Then chew the gum for two or three minutes (or choose a volunteer) and make three or four post-chewing observations, one of which should include the mass. Add these observations to the list on the board. Explain to students that they will be conducting an investigation of at least two types of gum using similar qualitative and quantitative observations and making inferences based on those observations.

Process Skill Activity

Students will work individually on this activity. Invite them to select at least two types of gum to complete Parts One and Two of the activity. Students may work in pairs to complete Part Three. Have students share the results of their investigations with the class.

▼ In this activity, students will make qualitative and quantitative observations of chewing gum and inferences based on those observations. In addition, students will design and carry out an investigation to compare different types of chewing gums.

Part Three Answer Key (answers may vary)

Gum as Food

1. mass
2. sugar
3. food

Gum Packaging

1. fresh
2. closely
3. wrappers
4. flat
5. sticking

Name _____ Date _____

Sweet but Sticky Observations and Inferences

In "Mysterious Journeys in the Life of a Raisin," you made only qualitative observations. In this activity, you will make both qualitative and quantitative observations.

What are qualitative observations? _____

_____

What are quantitative observations? _____

_____

PART ONE

There are many types of gum. Some gums are soft and others have a hard outer coating. Some are made with sugar and others are made with artificial sweeteners. Perhaps some gums will stretch farther than others after being chewed. Plan an investigation of different chewing gums. Write your plan, carry it out, and record your observations in the spaces below.

Investigation Plan

_____

_____

_____

_____

Qualitative Observations

_____

_____

_____

_____

Quantitative Observations

_____

_____

_____

_____

Teaching Science Process Skills © 1995 Good Apple

Name _____ Date _____

PART TWO ▶ Make inferences explaining your observations about different chewing gums by completing the following statements.

Chewed gum has less mass than unchewed gum because _____

_____

Chewed gum loses its flavor because _____

_____

_____

Chewing gum must be considered a food because _____

_____

_____

A new stick of gum is covered by aluminum and paper because_____

_____

_____

A paper wrapper is used on the outside of the aluminum and paper because

_____

_____

_____

White, powdery crystals can be found on the surface of some gum types because

_____

_____

Some chewing gum is wrapped and packaged in strips because _____

_____

_____

Some sticks of gum are stamped with patterns or initials because _____

_____

_____

PART THREE ▶ The paragraphs below provide an explanation of the chewing and packaging of chewing gum. Write the missing word(s) or phrase(s) that will make each sentence correct.

The Chewing Gum Story

Gum as Food

Chewing gum is made with rubber, plastic, wax, sugars, and flavorings. Some people do not think that chewing gum is a food substance. However, when gum is chewed, it loses _____. This loss is probably

_____ and other substances. As the gum is chewed, these materials mix with saliva and then are swallowed. This also explains why chewing gum loses its taste. Chewing gum must be considered a _____.

Gum Packaging

Often the form of an object is related to its function or purpose. Chewing gum is placed in aluminum and paper to keep it soft and _____. It stays fresh because the wrapper keeps the gum from losing moisture. The outer paper wrapper helps keep the aluminum and paper _____ around the gum stick. Packaging gum in individual _____ makes it possible to store some of the product for future use. The _____ stick shape is convenient for storing and removing the sticks from the package. The powdered sugar or starch on each gum stick helps prevent the gum from _____ to the paper. It is clear that the design of gum and its packaging serve a purpose.

PART FOUR ▶ In your own words explain what you have learned about chewing gum.

_____

_____

_____

_____

_____

Teaching Science Process Skills © 1995 Good Apple

Creating a Scientific Model

Teacher Notes

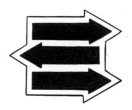

Materials (per group)

- sealed box containing a variety of items
- introductory information on creating a scientific model on page 64
- worksheet on page 65

Active Engagement Before the Task

Distribute the introductory information to each student. Individually, in small groups, or as a class read and discuss the material. Actively engage students' minds in preparation for the process skill activity by asking the following questions.

- Can you think of some scientific situations in which a model would be helpful to make observations and inferences?

- Can you think of some situations in your own life when a model would have been helpful to you?

- How is it possible for scientists to create models of things they cannot even see?

In this activity, students will practice creating a scientific model by making observations and inferences about a special box or boxes you have constructed.

Process Skill Activity

Prepare a "mysterious black box" for each group before class by placing some items that will make interesting sounds inside a sealed shoe box or any box of similar size. Some of the items may be attached to other items or to the inside of the box if you wish. (If your class is large or time is limited, one box for the entire class will suffice.) Cover the boxes with black paper. Students will work in groups to find out what is inside their box by first generating questions that can only be answered with yes or no. Using the same types of questions, they will try to ascertain if the components or variables inside the box are put together in any special way. They will make observations, inferences, and draw a model of the interior of the box. Once the class has shared their inferences, boxes may be opened for comparison with the drawings.

Creating a Scientific Model

Sometimes scientists make inferences about an object that is hard to see because it is very far away, very large, or very small. Sometimes scientists make inferences about things they are not able to see at all. When many observations and inferences are needed, it is often easier to create a scientific model. A model is a tool scientists use to help them understand a set of observations and make better inferences. Models are frequently written or drawn by scientists. Often they are generated by computers. Just like scientists, you have probably used models to help you understand inferences and ideas. For example, when you need help finding your way in an unfamiliar part of town, you use a model called a map.

The ability to create models is a skill you can use to good advantage in many situations—not only in science.

Teaching Science Process Skills © 1995 Good Apple

Creating a Scientific Model

Try to recreate a scientific model. You will see a "mysterious black box" with some inexplicable sounds. Ask questions that can be answered with a yes or a no. The purpose of your questions is to find out what is inside the box and to find out if the components, or variables, in the box are put together in any special way.

Write your observations and inferences in the spaces below. Then draw a model of the interior of the box.

Observations

_____

_____

_____

Inferences

_____

_____

_____

_____

Model

Inferences Are Risky!

Materials (per student)

- story on page 67
- worksheet on page 68

Active Engagement Before the Task

Students have learned about making inferences from their observations. They know that an inference is a rational explanation of an observation. However, sometimes inferences can explain observations and still be incorrect. This can occur because not all the necessary observations have been made. In other words, all the facts are not known. Actively engage students' minds in preparation for the activity by asking the following questions.

- Do you think an inference can explain an observation and still be incorrect?

- What might be some reasons for incorrect inferences?

- Which of your five senses do you think is the least likely to trick you? Why do you think so?

In this activity, students will hear a story that will help them understand the risk of making inferences without all the facts.

Process Skill Activity

Students will work individually. Read the following story aloud to your class. Distribute the worksheet and have students answer the questions concerning the story and their study of observations and inferences. Lead the class in a discussion on their thoughts about the story. Review their understanding of the skills.

Elephant Observations

Long ago in a distant land, six blind men lived together. All of them had heard of elephants, but they had never "seen" one. When they heard that an elephant and his trainer would be visiting their village, they all wanted an encounter with this beast. They made their way to the site where the elephant was being kept. Each blind man touched the elephant and made his observations. The observations are listed below.

One man touched the elephant's side and said,

"An elephant is like a wall."

Another man touched the trunk and said,

"An elephant is like a snake."

Another man touched a tusk and said,

"An elephant is like a spear."

Another man touched a leg and said,

"An elephant is like a tree."

Another man touched an ear and said,

"An elephant is like a fan."

The last man touched the tail and said,

"An elephant is like a rope."

Seven Blind Mice by Ed Young (Philomel, 1992) offers another delightful rendition of this tale.

Inferences Are Risky!

▼ Did the blind men make appropriate inferences? Explain.

_____

_____

▼ How might the blind men improve their inferences?

_____

_____

▼ One of the characteristics of science is that scientists communicate their ideas, observations, results, and inferences with each other. Why is this a good idea?

_____

_____

_____

You have spent a number of days studying qualitative and quantitative observations. You have practiced making measurements using the metric system and you have worked on the skill of inferring. In the space below, write a sentence or two explaining what you have learned.

Qualitative Observations

_____

_____

Quantitative Observations

_____

_____

Measurement

_____

_____

Inferences

_____

_____

Teaching Science Process Skills © 1995 Good Apple

*I*dentifying and Manipulating Variables

O b j e c t i v e s

Based on experiences with variables, research questions, and additional discussion, students will be able to:

▼ Define or select the definition for *variable*.

▼ Explain the role of variables in the process of science.

▼ Successfully identify a list of potential variables in any given event

▼ Define or select the definitions for the three types of variables.

▼ Identify the manipulated, responding, and controlled variables in a given event.

▼ Select the most appropriate operational definition for a variable from a given list.

▼ Explain the role of an operational definition in the process of science.

▼ Operationally define a given variable.

▼ Define or select the definition for a research question.

▼ Explain the role of a research question in the scientific process.

▼ Identify the important components of an appropriate research question.

▼ Write a research question for a given variable or for a relationship between variables.

▼ Given an event, successfully identify possible variables, write appropriate research questions, and make effective operational definitions for those variables.

Identifying and Manipulating Variables

By studying simple actions and reactions, such as how raisins act in a baking soda solution, you have learned that observing and inferring are the basis of science. But actions and reactions in the natural world are often complex. Sometimes they are so large (like the explosion of a volcano), or so small (like the movement of a Euglena), or so distant (like the birth of a star), or so spread over time (like the movement of a glacier) that it is impossible for the human mind to understand them in their entirety.

The scientific approach to understanding such events is a process that breaks complex events into parts that can be studied and understood. These parts of an event or system are called *variables*. Variables are factors, conditions, and/or relationships that can change or be changed in an event or system. In order to learn about scientific investigation, you first need to learn the skill associated with identifying and manipulating variables.

Teaching Science Process Skills © 1995 Good Apple

Identifying Variables

Materials (per student)

- introductory information on identifying and manipulating variables on page 70
- worksheet on pages 72-73

Active Engagement Before the Task

Distribute the introductory information to each student. Individually, in small groups, or as a class read and discuss the material. Explain that a variable is a factor or condition that affects a system, event, or process in some way. For example, the likelihood of rain can be affected by variables, such as air temperature, air pressure, wind speed, cloud conditions, or altitude. Emphasize that variables are important in science because they are part of a method that allows scientists to make complexity simple. Actively engage students' minds in preparation for the process skill activity by asking the following questions.

- What variables might affect the taste of a pizza?

- What variables might affect the miles per gallon of an automobile?

- What variables might affect the score of an archer?

Process Skill Activity

Students will work in groups to identify four variables for a number of situations. Once each group has completed working, invite them to share with the entire class. Post their responses to the question, "What have you learned about variables?"

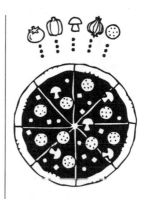

In this activity, students will identify possible variables for a variety of situations.

Name _____ Date _____

Identifying Variables

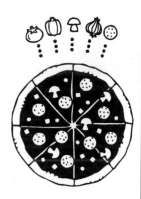

Identify possible variables in the following situations.

What variables can affect the sprouting of a bean seed?

Variable 1: _____

Variable 2: _____

Variable 3: _____

Variable 4: _____

What variables can affect the number of books sold by a door-to-door salesperson?

Variable 1: _____

Variable 2: _____

Variable 3: _____

Variable 4: _____

What variables can affect the number of fish in a lake?

Variable 1: _____

Variable 2: _____

Variable 3: _____

Variable 4: _____

What variables can affect attendance at a football game?

Variable 1: _____

Variable 2: _____

Variable 3: _____

Variable 4: _____

What variables can affect the number of eggs laid by a chicken?

Variable 1: _____

Variable 2: _____

Teaching Science Process Skills © 1995 Good Apple

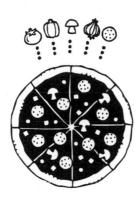

Variable 3: _____

Variable 4: _____

What variables can affect the taste of a soft drink?

Variable 1: _____

Variable 2: _____

Variable 3: _____

Variable 4: _____

What variables can affect the amount of fruit produced by an apple tree?

Variable 1: _____

Variable 2: _____

Variable 3: _____

Variable 4: _____

What variables can affect the speed of a runner in a 100-yard dash?

Variable 1: _____

Variable 2: _____

Variable 3: _____

Variable 4: _____

What have you learned about variables?

_____

_____

_____

_____

_____

_____

Defining Variables Operationally

Materials (per student)

- explanatory information on operational definitions on page 75
- worksheet on pages 76-77

Active Engagement Before the Task

Distribute the explanatory information on operational definitions. Individually, in small groups, or as a class read and discuss the material. Actively engage students' minds in preparation for the process skill activity by asking the following questions.

- A teacher is interested in investigating the effect of homework on test results. What are two operational definitions for the variable "homework?"

- A shopkeeper wants to find out if window posters affect sales. Give two operational definitions of the variable "window posters."

- A student wants to measure which pizza toppings her friends prefer. What is an operational definition of the variable "pizza topping preference?"

▼ In this activity, students will become familiar with operational definitions through definition, example, and practice.

Process Skill Activity

Students will work individually or in pairs to complete the worksheet. You may wish to have three students or groups carry out the paper towel absorbency tests to see if the same towel wins under all three conditions. Results may be shared with the class.

What is an Operational Definition?

One of the important decisions a scientist must make is to determine how measurement of the variable will be made. The method used to measure a variable is called an *operational definition*. An operational definition indicates the way a measurement will be performed. Once a scientist has decided on a method, that method must be reported to other scientists, so they also can test the investigation results. Any scientist can read an operational definition and easily understand or perform the same measurement. The examples below show operational definitions of variables.

Example One
A student wants to test the effects of vitamin C on the health of students in her science class. The variable "health of students" could be defined in the following ways.

- the number of colds experienced during a month

- the number of days absent due to sickness in a month

- the number of people with coughs in a month

Example Two
A student wants to test the effect of "Don't Litter" posters on the trash problem at his school. The variable "trash problem" could be defined in the following ways.

- the number of candy wrappers on the playground

- the number of bags of trash collected

- the number of aluminum cans in the courtyard

Defining Variables Operationally

Your task is to think of operational definitions that might be used to measure variables in several situations. Before you begin, let's look at an example.

A student wants to measure the absorbency of paper towels, so absorbency is the variable. The student must create an operational definition for measuring the absorbency of paper towels. He develops three possible operational definitions.

- **The Dunk:** Measure the amount of water that remains after a crumpled paper towel has been placed in 25 ml of water for five minutes.

- **The Pour:** Measure the amount of water that collects after 25 ml of water has been poured through a crumpled paper towel.

- **The Lift:** Measure the height that water reaches after the end of a folded paper towel has been inserted in water for 15 minutes.

 PART ONE Think of operational definitions that might be used to measure variables in the following situations.

1. A student is interested in magnets. He wants to measure the strength of his favorite magnet.

Operational definition of the variable "magnet strength"

_____

2. A student is interested in investigating the germination (sprouting) of seeds.

Operational definition of the variable "germination"

_____

3. A student wants to measure which soft drink her classmates prefer.

Operational definition of the variable "soft drink preference"

_____

4. A student wants to find out how interested her classmates are in reading books about science.

Operational definition of the variable "interest in reading books about science"

_____

Teaching Science Process Skills © 1995 Good Apple

5. A student wants to find out if study affects science grades.

Operational definition of the variable "study"

_____

Operational definition of the variable "science grade"

_____

 PART TWO Tho following investigation contains operational definitions for a variable. Identify the variable and the operational definitions for the variable.

A study was done to determine the effect of distance running on breathing rate. Students ran different distances and the rate of breathing was measured. One group ran ¼ km, a second group ran ½ km, and a third group ran 1 km. Immediately after running, breathing rate was checked by counting the number of breaths taken in one minute.

Variable_____

_____

_____

Operational definition _____

_____

_____

Operational definition _____

_____

_____

_____

Helicopter Happening

Materials (per student)

- scissors
- ruler
- worksheet on page 79
- helicopter pattern on page 80

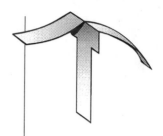

Active Engagement Before the Task

Review the definition of variables and of operational definitions of variables. Actively engage students' minds in preparation for the process skill activity by asking the following questions.

- Can an object without moving parts be made to rotate other than by physically spinning it? Explain your answer.

- What might affect an object's ability to spin?

- What would you say is the single most important element in an object's ability to spin? Explain your choice.

Process Skill Activity

Students can work individually to prepare the rotating object but may want to work in groups to discuss variables and operational definitions. Emphasize the importance of accuracy in cutting and folding. Encourage groups to share their observations and results with the class.

In this activity, students will make thoughtful observations and inferences about how a rotating object (paper helicopter) works. They will also identify and operationally define variables that might affect the action of the helicopter.

Helicopter Happening

You know that variables are factors that can affect an event and you know how to define variables operationally. This activity will give you more practice in identifying and operationally defining variables.

Carefully cut out the pattern for the rotating object and follow the assembly directions. Test the device to find out how it works.

▼ Record your observations and inferences.

_____

_____

_____

_____

_____

▼ What are some possible variables that could affect how it flies?

_____

_____

_____

_____

▼ Identify an operational definition for each variable you listed above.

_____

_____

_____

_____

_____

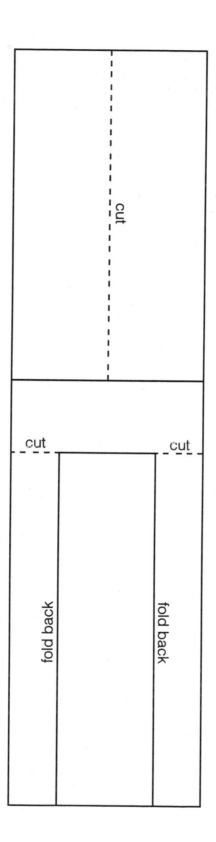

cut

cut   cut

fold back   fold back

Telephone Technology Made Simple

Teacher Notes

Materials (per group)

- string or twine
- 2 cans or cups
- ice pick or screw driver
- worksheet on page 82

Active Engagement Before the Task

Review how to identify and operationally define variables. Actively engage students' minds in preparation for the process skill activity by asking the following questions.

- How many of you have ever made or used a string telephone?

- What problems did you encounter with your telephone?

- What do you think the maximum distance would be for a string phone to be effective?

Process Skill Activity

You may wish to have students make string telephones. They require two cans or cups, string, and a sharp object (ice pick or screw driver) to poke a hole in each can. Provide a variety of materials so that products will differ. Once groups have identified variables and operational definitions on the worksheet, encourage them to share results with the class. If string phones have been constructed, try them out and bring the class to consensus on the most effective items for string telephone construction.

In this activity, students will identify and operationally define variables that might affect a string telephone.

Name _____ Date _____

Telephone Technology Made Simple

Here is another chance to practice identifying variables and creating operational definitions for those variables. Think about a string telephone. Perhaps you have made or used one. What parts or variables could be changed?

List the variables and provide an operational definition that would define or measure each variable.

Variable 1 _____

Operational Definition _____

_____

Variable 2 _____

Operational Definition _____

_____

Variable 3 _____

Operational Definition _____

_____

Variable 4 _____

Operational Definition _____

_____

Three Kinds of Variables

Materials (per student)

• worksheet on pages 84-86

Active Engagement Before the Task

Distribute the worksheet. Read the introductory information and explain the three kinds of variables. As an example, relate to students the experiment in which a young man wanted to find out whether heavy metal music, classical music, or no music at all would be most conducive to learning and, therefore, the best background music for doing homework. He used a hamster maze and three hamsters in his experiment. Actively engage students' minds in preparation for the process skill activity by asking the following questions.

• What would you predict was the result of his experiment and why? (classical music was the winner)

• What were the manipulated, responding, and controlled variables in this experiment?

• What possible fallacy might you cite in this experiment?

Process Skill Activity

Students may work individually or in pairs to identify the three types of variables. Point out the information on several controlled variables and the importance of manipulating only one at a time. Bring the class to consensus by sharing the results of student work.

 In this activity students will identify manipulated, responding, and controlled variables in several experiments.

Name _____   Date _____

Three Kinds of Variables

In a scientific investigation there are three kinds of variables. A *manipulated variable* (sometimes called the independent variable) is a factor or condition that is intentionally changed by an investigator in an experiment. A *responding or dependent variable* is a factor or condition that might be affected as a result of that change. A variable that is not changed is called a *controlled variable*. Consider the example below.

A student wanted to test how the mass of a paper airplane affected the distance it would fly. Paper clips were added before each test flight. As each paper clip was added, the plane was tested to determine how far it would fly. The mass of the plane (number of paper clips added) was the manipulated variable. The responding variable was the distance flown. A controlled variable in the experiment was the fact that the same plane was used for each trial.

For each experiment below, specify the manipulated, responding, and controlled variables.

▶ Two groups of students were tested to compare their speed working math problems. Each group was given the same problems. One group used calculators and the other group computed without calculators.

Manipulated variable_____

Responding variable _____

Controlled variable_____

▶ Students of different ages were given the same puzzle to assemble. The puzzle assembly time was measured.

Manipulated variable_____

Responding variable _____

Controlled variable_____

Teaching Science Process Skills © 1995 Good Apple

▶ A study was done to find if different tire treads affect the braking distance of a car.

Manipulated variable_____

Responding variable _____

Controlled variable_____

There can be several controlled variables. If an experiment is to be useful, only one variable at a time can be manipulated intentionally. All other variables must be controlled throughout all parts of the experiment. If more than one variable is altered, the results of an experiment cannot be interpreted with any validity.

Here is some practice with experiments having more than one controlled variable.

▶ An experiment was performed to determine how the amount of coffee grounds could affect the taste of coffee. The same kind of coffee, the same percolator, the same amount and type of water, the same perking time, and the same electrical source were used.

Manipulated variable_____

Responding variable _____

Controlled variable_____

Controlled variable_____

Controlled variable_____

▶ A study was done with an electromagnet system made from a battery and wire wrapped around a nail. Different sizes of nails were used and the number of paper clips that the electromagnet could pick up was measured.

Manipulated variable_____

Responding variable _____

Controlled variable_____

Controlled variable_____

Controlled variable_____

Name _____ Date _____

▶ A study was attempted to find if the length of the string in a string telephone affected its sound clarity.

Manipulated variable _____

Responding variable _____

Controlled variable _____

Controlled variable _____

Controlled variable _____

Controlled variable _____

In your own words, define manipulated, responding, and controlled variables.

_____

_____

_____

_____

_____

_____

_____

Teaching Science Process Skills © 1995 Good Apple

Using Variables in Research Questioning

Teacher Notes

Materials (per student)

- introductory information on using variables in research questioning on page 88
- worksheet on pages 89-90

Active Engagement Before the Task

Distribute the introductory information to each student. Individually, in small groups, or as a class read and discuss the material. Be sure students understand the difference between a *type one* and a *type two* question. Carefully review the rules for appropriate questioning. Actively engage students' minds in preparation for the process skill activity by asking the following questions.

- Can you think of another research question with regard to the paper helicopters? Is your question a *type one* or a *type two*?

- How can identifying variables help find a problem to investigate?

- What's your opinion? "Do men or women drive better?" Justify your opinion.

Process Skill Activity

Students can work in pairs or groups on this activity. Distribute the worksheet and introduce students to the variable wheel. Give them an opportunity to add some variables to the wheel and ask groups to share their variables.

*In this activity, students will learn to focus an investigation through research questions. They will also learn to distinguish between **type one** and **type two** research questions and practice writing both types.*

Using Variables in Research Questioning

Science uses investigations that require a problem to solve or answer. The most important part of every investigation is variables. When an investigator identifies the variables of an event, an interesting and important research question will often become obvious. A research question defines the problem to be studied. Once research questions have been formed, they influence the decisions that will be made about the subsequent investigation.

There are two types of research questions. A *type one* research question focuses on one variable. A *type two* research question focuses on a relationship between two variables—how one variable might affect a second variable.

Consider the paper helicopter you made earlier.

A *type one* research question might be, "How many rotations does a paper helicopter complete as it falls?" This question focuses on one variable—the number of rotations.

A *type two* research question could be, "To what extent does a paper helicopter's blade length affect its rotation?" This question asks about a relationship between two variables—the effect of blade length on rotation.

Here are some rules for writing appropriate research questions.

- State research questions in question form.

- Avoid questions that can be answered with a yes or no.

- Start research questions with phrases like, "To what extent . . ." or "What evidence indicates . . ."

- Include information, such as population (target group on which an investigation will focus) and area (geographical region in which an investigation will take place) that will limit the research question. For example, "To what extent does batting practice affect batting average?" is too broad. "To what extent does batting practice affect the batting average of Lincoln Middle School baseball players?" is more specific and, therefore, a better research question.

Teaching Science Process Skills © 1995 Good Apple

Using Variables in Research Questioning

You will have several opportunities to practice writing research questions using the variable wheel below. The wheel is a scientific model to help you identify variables and write research questions. Here's how it works. Let's assume that you want to study baseball batting averages. The operational definition is how frequently a pitched ball is hit safely. The variable "batting average" is in the center of the wheel. Other variables that might affect batting averages are on spokes of the wheel. Write four more possible variables on the empty spokes. The wheel should help you see relationships between the manipulated variables and the responding variable "batting average."

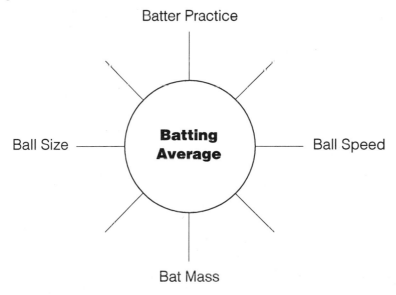

Try some *type one* research questions. Here's an example.

How much batting practice do Lincoln Middle School baseball players receive?

This research question focuses on only one variable—the amount of batting practice. Try writing two more *type one* questions in the space below.

_____

_____

_____

_____

Name _____ Date _____

Use the variables on the spokes of the wheel to write some *type two* research questions. Here's an example.

How does ball size affect the batting average of Lincoln Middle School baseball players?

_____

_____

_____

_____

The following research questions have not been written correctly. Refer to the rules for appropriate research questions and revise the questions.

1. Does mass make a paper helicopter rotate more?

Revision_____

_____

_____

2. Will longer strings change the sound in string telephones?

Revision_____

_____

_____

3. Do girls or boys hit a pitched ball more often?

Revision_____

_____

_____

Teaching Science Process Skills © 1995 Good Apple

Predicting

Objectives

Based on the experiences with this chapter and additional discussion, students will be able to:

▼ Define (or select the definition for) a prediction.

▼ Define the two kinds of predictions.

▼ Use data to make both interpolated and extrapolated predictions.

▼ Compare and contrast interpolated and extrapolated predictions.

▼ Examine graphed data to determine if a given prediction is interpolated or extrapolated.

▼ An important skill in science is predicting—the forecasting of future events. To understand predicting, it is important to remember that science is based on several assumptions or beliefs about the natural world. Scientists believe that there are cause-and-effect relationships in the natural world that control the world in a somewhat orderly manner. For example, predators (mountain lions) can cause a decrease in a prey (rabbits) population. This is a cause-and-effect relationship. Each time an apple releases from its branch, it will fall toward the center of the earth, regardless of the type of apple or its location on earth, because of orderly forces at work. The belief in cause-and-effect and orderly forces leads to the assumption that events in the natural world can be predicted.

However, some events are more accurately predicted than others. Predictions are based on past observations or available data. The amount of data available and the accuracy of the data can have a profound effect on the accuracy of the prediction. Eclipses and planet location, for example, can be predicted to the minute of occurrence. Predictions of weather or population changes, however, cannot be made as accurately. The assumption that the world behaves in an orderly manner helps scientists use available and accurate data to forecast future events.

Teaching Science Process Skills © 1995 Good Apple

Penny Predictions

Teacher Notes

Materials (per student)

- 1 penny, 1 nickel, 1 dime, 1 quarter
- water
- medicine dropper
- paper towels
- introductory information on predicting on page 92
- worksheet on page 94

Active Engagement Before the Task

Distribute the introductory information to each student. Individually, in small groups, or as a class read and discuss the material. Actively engage students' minds in preparation for the process skill activity by asking the following questions.

- Meteorology is a profession based almost entirely on prediction. Can you give an example of another profession that is similarly based on prediction?

- Can you give an example of a prediction that is easily made based on past observation and available data?

- What would be an example of a prediction which would be difficult to make based on past observation and available data?

Process Skill Activity

Students will work individually on this activity. Remind them that a previous activity, "It's in the Liquid," might give them some food for thought as they make their predictions. Once the task has been completed, invite students to share their responses with the class.

In this activity, students will predict and test how many drops of water various coins will hold.

Name _____ Date _____

Penny Predictions

You will be placing drops of water on various coins with a medicine dropper. First you will predict how many drops each coin can hold. Then you will count the number of drops each coin holds before the water runs off.

Research Question: To what extent does the size of coins affect the number of drops of water the coin will hold?

Manipulated Variable _____

Responding Variable _____

Controlled Variable _____

Controlled Variable _____

1. Predict how many drops of water the penny will hold. Record your prediction on the table below.

2. Drop water on the penny. Record how many drops the penny held.

3. Dry the penny and repeat step 2 two more times. Record your findings below.

4. Average your three trials for a more accurate idea of how many drops of water a penny will hold.

5. Do the investigation again using different coins. Record your results below.

Type of Coin	Prediction	Trial #1	Trial #2	Trial #3	Average
penny					
nickel					
dime					
quarter					

Discuss your thoughts about predicting. Did your ability to predict become more accurate as you performed the investigation with coins other than the penny?

_____

_____

_____

_____

Teaching Science Process Skills © 1995 Good Apple

Predictions about Paper and Plastic

Materials (per student)

Cut squares of each of the first seven materials listed below as well as some random choices of your own. Place the squares in appropriately labeled bags for easy student access.

- newspaper
- plastic bag
- grocery bag
- produce bag
- napkin
- copier or typing paper
- construction paper
- oil
- dropper
- newspaper, blotter paper, or paper towels
- worksheet on page 96

In this activity, students will predict and test the ability of oil to pass through various types of paper and plastic.

Active Engagement Before the Task

Review the information on predictions. Actively engage students' minds in preparation for the process skill activity by asking the following questions.

- Based on your own past observations, in what material would you wrap a bacon, lettuce and tomato sandwich with lots of mayonnaise if you had to put it in your pocket? Why did you make this choice?

- Why do you think it's important for some wrapping materials to allow air to pass through?

- Which do you think would make a more effective napkin—absorbent or non-absorbent material? Why do you think so?

Process Skill Activity

Students may work individually, in pairs, or in small groups on this activity. Encourage them to choose three or four additional materials to test and to add those to the chart. Protecting the work surface with absorbent material will help students determine if and when oil has passed through the material they are testing. Have students make all their predictions before they begin the actual testing. Once testing is complete, students may add qualitative or quantitative observations. Invite students to share their findings with the class. Bring the class to consensus on the results of the experiment.

Name _____ Date _____

$\mathcal{P}$redictions about Paper and Plastic

Predict and test the ability of oil to pass through various types of paper and plastic.

Research Question: To what extent will oil pass through paper better than plastic?

Manipulated Variable: plastic and paper materials

Responding Variable: drops of oil

Qualitative Observations

_____

_____

Select three or four additional substances to test and write them on the chart. Make all your predictions first. Make some qualitative observations. Then go back and test each substance. If oil appears on the protective covering of your work surface, it has passed through the material you are testing. Check after each drop to ascertain how quickly it moves through the manipulated variable. Conclude with some quantitative observations.

Substance	Predictions Will oil pass through?		Number of Drops
	Yes	No	Number or N/A
newspaper			
grocery bag			
produce bag			
napkin			
copier paper			
construction paper			
plastic bag			

Quantitative Observations_____

_____

Predictions about Insects

Teacher Notes

Materials

- split peas
- yellow lentils
- black beans
- pinto beans
- lima beans

- navy beans
- black-eyed peas
- grassy/sandy area in the schoolyard
- small plastic cups
- worksheet on page 98

Active Engagement Before the Task

Locate an area for this activity that has both grass and dirt or sand. Review qualitative and quantitative observations as well as infer ences with students. Explain that they will use various types of beans and lentils to represent insects and that they will be acting as foraging birds. Actively engage students' minds in preparation for the process skill activity by asking the following questions.

- Birds have the best vision of all classes of animals. How might one deduce that birds have excellent color acuity as well?

- Can you give some examples of protective coloration in insects?

- Do you think birds choose insects to eat because they are easily visible or because they are tasty? Why do you think so? How might you find out?

Process Skill Activity

Students will work individually or in small groups on this activity. Give each student a small plastic cup containing at least one of each type of bean or pea you are using. Identify each one by name so students can use the chart correctly. Have students make predictions on the worksheet about the ease of finding each one. Encourage them to make some qualitative observations about the peas and beans at this point. Students should then empty their cups into a central supply which you will scatter around the outdoor area. Students will take their empty cups out for the "insects" they find. At the end of the allotted time, bring your students in to make some quantitative observations and inferences about their results. Small groups may share inferences and come to consensus before sharing with the class. Bring the class to consensus on the results of the activity.

In this activity, students will make predictions, observations, and inferences about the effect of insect size and color on the insect's likelihood of being found and eaten by a bird.

Name _____ Date _____

Predictions about Insects

Research Question: Will the size and color of an insect protect it from being found by a bird?

Manipulated Variable: size and color of insects

Responding Variable: number of insects found

"Insect"	Predictions Easy to find?		Number Found
	Yes	No	
split peas			
yellow lentils			
black beans			
pinto beans			
lima beans			
navy beans			
black-eyed peas			

Qualitative Observations

_____

_____

Quantitative Observations

_____

_____

Inferences

_____

_____

Teaching Science Process Skills © 1995 Good Apple

Hypothesizing

Objectives

Based on the experiences with this chapter and additional discussion, students will be able to:

▼ Identify a hypothesis from a given list of statements.

▼ Explain why variables are important in the process of hypothesizing.

▼ Write a hypothesis using two variables.

▼ Define (or select a definition for) a hypothesis.

▼ Compare and contrast a hypothesis with a research question.

▼ Explain the relationship between an inference and a hypothesis.

*H*ypothesizing

You have learned that variables are important not only in writing research questions but also in making predictions. Predicting is the process of using observations or data along with other kinds of scientific knowledge to forecast future events or relationships. A hypothesis is a special kind of prediction that forecasts how one variable will affect a second variable. These variables are the manipulated variable, which is changed intentionally by the investigator, and the responding variable, which is observed or measured to determine if or how much it is affected. Hypotheses express a logical explanation that can be tested. Investigators find them useful because they specify an exact focus for an experiment. Here is an example of a hypothesis.

If the temperature of sea water increases, **then** the amount of salt that will dissolve in that water increases.

Water temperature and amount of dissolved salt are the variables used in this hypothesis. The investigator is predicting that warmer water will have more dissolved salt than colder water. An investigator can design an experiment that manipulates the temperature of several samples of water from the same source. Dissolved salt levels can then be measured in each sample. Analysis of the data would indicate the extent to which dissolved salt levels are related to water temperature.

Notice that the sample hypothesis is expressed as an "If . . . , then . . ." sentence. This form, while not always necessary, is a helpful way to learn to write a hypothesis. Here are two examples.

Example One
Some students want to find out what kind of pizza is preferred by their classmates.

Teaching Science Process Skills © 1995 Good Apple

Research Question
What kind of evidence indicates that middle school students prefer one kind of pizza more than another? (*type one* question)

Hypothesis
If middle school students are questioned about pizza preference, they will prefer pepperoni over cheese pizza.

Example Two

An investigator has observed that chickens lay more eggs at certain times of the year. It has also been observed that this occurs during the late spring and summer months. An inference has been made that the extra eggs are due to longer daylight hours. Amount of daylight and chicken egg production are the variables the investigator has decided to investigate.

Research Question
To what extent does the length of daylight affect chicken egg production? (*type two* question)

Hypothesis
If the length of daylight increases, then chicken egg production will increase.

Steps for Writing a Good Hypothesis

- Identify variables in a given event or relationship.

- Identify a pair of variables that might be logically related.

- Identify the manipulated and responding variables.

- Write the hypothesis using the following format.

If the [manipulated variable] increases or decreases, then the [responding variable] will increase or decrease.

Writing Hypotheses

Teacher Notes

Materials (per student)

- introductory information on hypothesizing on page 100-101
- worksheet on page 103-104

Active Engagement Before the Task

Distribute the introductory information on hypothesizing to each student. Individually, in small groups, or as a class read and discuss the information. Actively engage students' minds in preparation for the process skill activity by asking the following questions.

- What's wrong with the following hypothesis—cold pickles are crispier?

- How might you restate the pickle information in a correct hypothesis?

- What is the difference between a *type two* research question and a hypothesis?

Process Skill Activity

Students may work individually or with a partner on this activity. Once students have finished Part One, stop and discuss their hypotheses. Ask a volunteer to state an original hypothesis. Have another student identify the manipulated and responding variables in the original hypothesis. Then encourage students to finish the activity by writing four original hypotheses and identifying the variables. Invite them to share these in small groups or with the class.

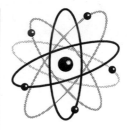

In this activity, students will practice writing hypotheses when given the manipulated and responding variables.

Name _____ Date _____

Writing Hypotheses

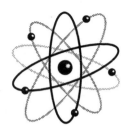

 PART ONE Write hypotheses for the following.

1. Manipulated variable: length of paper helicopter blades

Responding variable: rotational speed

Hypothesis _____

_____

2. Manipulated variable: length of string tolophone

Responding variable: clarity of sound

Hypothesis _____

_____

3. Manipulated variable: baseball batting practice

Responding variable: batting average

Hypothesis _____

_____

4. Manipulated variable: amount of gas bubbles

Responding variable: number of raisin round-trips

Hypothesis _____

_____

5. Manipulated variable: temperature of solution

Responding variable: dissolving time of powdered drink mix

Hypothesis _____

_____

6. Manipulated variable: depth of Lake Conroe

Responding variable: water temperature

Hypothesis _____

_____

Name _____ Date _____

7. Manipulated variable: number of recycling posters

Responding variable: amount of aluminum cans in courtyard

Hypothesis _____

_____

 Try writing your own variables and hypotheses.

1. Manipulated variable _____

Responding variable _____

Hypothesis _____

_____

2. Manipulated variable _____

Responding variable _____

Hypothesis _____

_____

3. Manipulated variable _____

Responding variable _____

Hypothesis _____

_____

4. Manipulated variable _____

Responding variable _____

Hypothesis _____

_____

Teaching Science Process Skills © 1995 Good Apple

The Relationship between Observing, Inferring, and Hypothesizing

Materials (per student)

• worksheet on pages 106-107

Active Engagement Before the Task

Explain to students that all the skills they have learned so far—observing, inferring, identifying variables, and hypothesizing—fit together to form a basis for scientific investigation. Distribute the worksheet. Review and discuss the process of examining the observation about golden retrievers and dachshunds. Actively engage students' minds in preparation for the process skill activity by asking the following questions.

• Here's an observation. A wooden boat had lots of barnacles. A fiberglass boat had relatively few barnacles. What possible explanations can you think of for this phenomenon? (Write inference #1, variables, and hypothesis #1 on the board based on the first student explanation.)

• Can anyone come up with another set of options? (Write inference #2, variables, and hypothesis #2 based on the second student explanation.)

• Which hypothesis would be easiest to test?

Process Skill Activity

Students will work individually or in pairs to create an original situation. They will identify observations, inferences, variables, and hypotheses. There should be at least two complete sets of options for each situation. Once the task has been completed, invite students to share their work for discussion.

▼ In this activity, students will learn how the skills of observing, hypothesizing, and inferring fit together. They will then create an original story in which they identify observations, inferences, variables, and hypotheses.

Name _____ Date _____

The Relationship between Observing, Inferring, and Hypothesizing

You have learned about observing, inferring, identifying variables, and hypothesizing. It's time to understand how these process skills fit together.

Here's an example of how observing leads to inferring, which in turns leads to identifying variables and hypothesizing.

Observations

A golden retriever had a lot of fleas. A dachshund had fewer fleas.

The following process examines these observations.

Option 1

Inference 1
Golden retrievers have more flea habitat than dachshunds.

Variables
Size of dog and flea population.

Hypothesis 1
If dogs are larger, then they will have a larger flea population than smaller dogs.

Option 2

Inference 2
Golden retrievers have longer hair than dachshunds.

Variables
Length of hair and flea population.

Hypothesis 2
If dogs have longer hair, then they will have more fleas than dogs with shorter hair.

Option 3

Inference 3
The golden retriever did not have a flea collar and the dachshund did.

Variables
Presence of flea collar and flea population.

Hypothesis 3
If dogs have flea collars, then they will have fewer fleas than dogs without flea collars.

Teaching Science Process Skills © 1995 Good Apple

Name _____ Date _____

▶ Create your own situation in the space below. Identify observations, inferences, variables, and hypotheses. List at least two complete sets of options (inferences, variables, and hypotheses) for your observations.

_____

_____

_____

_____

_____

_____

_____

_____

_____

_____

_____

_____

_____

_____

_____

_____

_____

_____

_____

_____

Hypothesizing about Discrepant Events

Teacher Notes

Materials (per group)

- introductory information on hypothesizing about discrepant events on page 110
- materials on worksheet lists on pages 111-114

Active Engagement Before the Task

(A day or two before you plan to begin studying discrepant events, set up the Blue Banana experiment on page 109.) Call students' attention to the color of the water in both jars and advise them to observe the water over the course of the next few days. To begin your study of discrepant events, distribute the introductory information to each student. Individually, in small groups, or as a class read and discuss the information. Actively engage students' minds for the process skill activities by asking the following questions as well as questions about the Blue Banana experiment.

- Both cloth bundles are full of garden dirt. What logical explanation can you think of for the color change in one jar?

- Demonstrate the phenomenon of the ping pong ball in the funnel. How can you explain the activity of the ping pong ball? Is there any way to blow it out of the funnel?

- What do you think would happen if you released the ping pong ball into an upward-directed stream of warm air from a hair dryer? (The ping pong ball will hover in mid-air over the hair dryer. Try it!)

In the following activities, students will observe a variety of discrepant events and practice making thoughtful observations, logical inferences, careful identification of variables, and hypotheses. Rather than having your entire class do every activity, consider assigning a different activity to each small group.

Process Skill Activities

Students will work in small groups. Each group may be assigned a different activity—choose from either Bobbing Cans, Some Think It's Hot, A Bag with Holes, or Paper Towel Roll-up. This would be an excellent opportunity for groups to make formal presentations to the class. These might include a visual and would certainly include their observations, inferences, variables, and hypotheses. Refer to the individual worksheets for the materials lists.

Bobbing Cans

Provide regular and diet soda. Due to the difference in mass of sweeteners used, regular soda sinks while diet drinks float.

Some Think It's Hot

Because metal conducts heat away from the fingers, the BB's will feel cold. The beans will not. However, temperatures in both containers will be the same.

A Bag with Holes

Air pressure keeps the water in the bag when it is zipped shut.

Paper Towel Roll-up

Use some plain and some patterned sheets of paper towel. Apply spray-on fabric protector to all the plain sheets. Alternate treated and untreated towels in a stack. Treated towels will not absorb water.

* *

Blue Banana

Use this interesting experiment to help introduce your students to discrepant events. Set it up a day or two before your introduction.

Materials
- banana slice
- garden soil
- water
- 2 large jars
- methylene blue
- 2 rubber bands
- 2 small pieces of cloth

Fill two large jars with water and add several drops of methylene blue to each. Place a slice of peeled banana and two to three tablespoons of garden soil on a small piece of cloth. Wrap the cloth into a ball, close it with a rubber band, and place the ball into one of the jars. In the second jar, place only a ball of soil wrapped in cloth. Students will observe a color change due to oxygen being used as the banana decomposes. The water in that jar will become clear.

Hypothesizing about Discrepant Events

▼ The unexpected often happens in the natural world. These unexpected events fly in the face of common sense or what we think we know. For example, when a ping pong ball is placed inside a funnel and air is blown up through the narrow stem, common sense would indicate that the ball would fly out from the force of moving air. But it doesn't. Another common unexpected event in the natural world is the abundance of marine fossils in deserts and on mountain tops. This indicates that at one point in geologic time these areas were covered by oceans. These types of unexpected events that have logical explanations are called *discrepancies* or *discrepant events*. A scientific investigator must take time for thoughtful observations, logical inferences, careful identification of variables, and hypothesizing.

Teaching Science Process Skills © 1995 Good Apple

Name _____ Date _____

Bobbing Cans

Predict and then test to see what will happen to two types of canned soda when placed in water.

Materials

• aquarium tank filled with water
• unopened diet and regular soda in cans

1. Predict what will happen when several cans of soda are placed in an aquarium filled with water.

_____

_____

_____

_____

2. Place the cans of soda in the aquarium.

3. Write observations, inferences, variables, and a hypothesis below.

Observations _____

_____

Inferences _____

_____

Variables _____

_____

Hypothesis:

If_____

_____

then_____

_____

Name _____ Date _____

Some Think It's Hot

Predict and then test to see if temperatures vary between a cup of BB's and a cup of beans.

Materials

- cup half full of BB's
- cup half full of beans
- thermometer

1. Predict which is colder—a cup half full of BB's or a cup half full of beans.

_____

2. Place one finger in each cup and record your observations.

_____

3. Measure the temperature of each cup. Record your observations.

_____

_____

4. Make inferences, identify variables, and write a hypothesis.

_____

_____

Observations _____

_____

Inferences _____

_____

Variables _____

_____

Hypothesis:

If_____

_____

then _____

_____

Teaching Science Process Skills © 1995 Good Apple

A Bag with Holes

Predict and then test to see what will happen when you poke several holes into a plastic bag filled with water.

Materials

- plastic bags with zipper closures
- water
- sharp pencils

1. Fill your bag with water and zip it closed. What do you think will happen when you poke a hole into it with a pencil?

_____

_____

_____

2. Poke holes into the bag with a pencil.

3. Record your observations, inferences, variables, and hypothesis.

Observations _____

_____

Inferences _____

_____

Variables _____

_____

Hypothesis:

If_____

_____

then _____

_____

_____

Name _____ Date _____

Paper Towel Roll-up

Predict and then test the absorbency of two paper towels.

Materials

- stack of paper towels
- 2 beakers of water

1. Take two paper towels from the stack. You should have a plain sheet and a patterned sheet. Which towel do you think will absorb more water? Why?

_____

_____

2. Roll each paper towel into a cylinder. Place one towel in Beaker A and the other in Beaker B. Record your observations, inferences, variables, and hypothesis below.

Observations _____

_____

Inferences _____

_____

Variables _____

_____

Hypothesis:

If_____

_____

then _____

_____

Teaching Science Process Skills © 1995 Good Apple

Organizing and Interpreting Data

Objectives

Based on the experiences with the activities in this chapter and subsequent discussion, the students will be expected to:

▼ Define (or select definitions for) the following—graph, data table, conclusions, inferences, recommendations.

▼ Graph both descriptive and continuous data.

▼ Complete a data interpretation for a given set of data, including data charts, graph, conclusions, inferences, hypothesis decision, and recommendations.

▼ Compare and contrast descriptive data and continuous data.

▼ Identify both appropriate and inappropriate interpretations drawn from a given set of data.

▼ Demonstrate the ability to identify and name basic trends between variables from graphed data.

▼ Identify and name the basic components of a graph.

Organizing and Interpreting Data

▼ Before conducting a meaningful investigation, it is important to learn how to organize the data you have collected. By organizing data, a scientist can more easily interpret what has been observed. Making sense of observations is called *data interpretation*. Since most of the data scientists collect is quantitative, data tables and charts are usually used to organize the information. Graphs are created from data tables. They allow the investigator to get a visual image of the observations which simplifies interpretation and drawing conclusions. Valid conclusions depend on good organization and clear interpretation of data.

Two types of graphs are typically used when organizing scientific data—bar graphs and line graphs. Descriptive data requires a bar graph. This is the type of data that comes from research questions asking about variables that will be counted. For example, "To what extent do the children in Sawmill School prefer tacos over pizza?" Continuous data requires a line graph. This type of data comes from research questions that ask about variables to be investigated over time. For example, "What evidence indicates that students at Old Turnpike School will gain weight if they eat pizza daily for a month?"

Since drawing conclusions is the final step of any investigation, tables, charts, and data interpretation are extremely important.

Teaching Science Process Skills © 1995 Good Apple

Data Three Ways

Teacher Notes

Materials (per student)

- introductory information on organizing and interpreting data on page 116
- worksheet on pages 118-119

Active Engagement Before the Task

Distribute the introductory information to each student. Individually, in small groups, or as a class read and discuss the material. Actively engage students' minds in preparation for the process skill activity by asking the following questions.

- What type of information would you expect to have found on charts kept by early man?

- Can you think of an activity you do at home that could be done more efficiently with the help of a chart or graph?

- Other than a scientist, what professionals would have use for graphs and charts?

Process Skill Activity

Students may work in pairs or small groups to discuss the first two data sets. Once they have determined the strengths and weaknesses of each, students can create a data table. Groups should come to consensus concerning the value of the table. Invite each group to share with the class and bring the class to consensus.

In this activity, students will compare data in three forms—narrative, pictorial, and charted—in order to assess the strengths and weaknesses of each form.

Name _____   Date _____

Data Three Ways

Compare the following sets of data. Then tell which data set communicates the information better. Give reasons for your choice.

Data Set Number One

Our sun has a surface temperature of about 5538°C. The innermost planet is Mercury. It has a surface temperature of about 327°C. The next planet, Venus, has a surface temperature of about 482°C. Our home planet Earth is next. Its surface temperature is about 14°C. Mars is the fourth planet and its surface temperature is about -23°C. Jupiter comes after Mars. Jupiter has a surface temperature of about -151°C. Saturn is next with a surface temperature of about -184°C. Uranus is after Saturn. Its surface temperature is about -207°C. Next is Neptune whose surface temperature is about -223°C. Pluto is the outermost planet. It is so far away from the sun that its surface temperature has not been measured but it is estimated to be about -230°C.

Data Set Number Two

Sun	Mercury	Venus	Earth	Mars	Jupiter	Saturn	Uranus	Neptune	Pluto
5538°C	327°C	482°C	14°C	-23°C	-151°C	-184°C	-207°C	-223°C	-230°C

Which set of data communicates information more easily?

_____

_____

What are the weaknesses and strengths of each?

_____

_____

_____

_____

Teaching Science Process Skills © 1995 Good Apple

Data Set Number Three

Both the paragraph of information and the pictorial representation presented the same data. Another way to show data is to use tables and charts.

Create a data table for the information presented about the planets.

Planet's Position from Sun	Surface Temperature
1.	
2.	
3.	
4.	
5.	
6.	
7.	
8.	
9.	

In the space below, discuss reasons why data tables are a good way for an investigator to present data for interpretation.

_____

_____

_____

_____

_____

_____

_____

Making Data Tables

Materials (per student)

• worksheet on pages 121-123

Active Engagement Before the Task

Review the advantages of data tables as discovered in the previous activity. Actively engage students' minds in preparation for the process skill activity by asking the following questions.

- What were the reasons for your finding that data tables were excellent methods of data presentation?

- Explain the difference between a manipulated and a responding variable.

- In the planetary data table, which was the manipulated and which was the responding variable?

Process Skill Activity

Distribute the worksheet and review the placement of manipulated and responding variables on a data table. Students may work individually or in small groups on this activity. Once data tables are complete, groups may compare results.

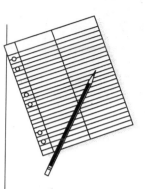

In this activity, students will make data tables using information from six student investigations.

Making Data Tables

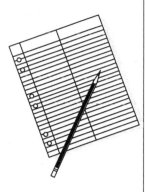

Make data tables for the data collected in the six investigations below. When making data tables, place the manipulated variable in the left column and the responding variable in the right column.

Investigation 1

A seed was planted. As the plant grew, it was measured over a six-day period.

day one - 0 centimeters, day two - 2 centimeters, day three - 5 centimeters, day four - 7 centimeters, day five - 8 centimeters, day six - 10 centimeters

Investigation 2

Tomato plants were grown at various temperatures. The number of tomatoes that grew on each plant was counted.

8°C - 4 tomatoes, 12°C - 10 tomatoes, 16°C - 14 tomatoes, 18°C - 18 tomatoes, 22°C - 24 tomatoes, 24°C - 16 tomatoes

Investigation 3

Different types of balls were bounced from a table top. The height of each bounce was measured.

golf ball - 54 centimeters, baseball - 9 centimeters, tennis ball - 48 centimeters, ping pong ball - 21 centimeters, Styrofoam ball - 3 centimeters

Investigation 4

Students in a science class were asked what type of pet they have. The number of each type of pet was recorded.

cats - 7, dogs - 6, hamsters - 3, hermit crabs - 2, snakes - 1

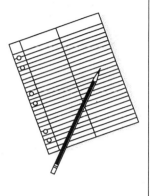

Investigation 5

The types and numbers of cars in the teachers' parking lot were recorded.

Ford - 12, Chevy - 16, Dodge - 9, Buick - 9, Volvo - 1, Honda - 2, Toyota - 1

Investigation 6

A student investigated and recorded how the amount of study time affected the scores on a science test.

0 hours - 58 points, 1 hour - 66 points, 2 hours - 82 points, 3 hours - 84 points, 4 hours - 88 points, 5 hours - 90 points, 6 hours - 88 points

Graphing

Teacher Notes

Materials (per student)

- introductory information on graphing procedures on pages 125-127
- worksheet on pages 128-131
- ruler

Active Engagement Before the Task

Distribute the introductory information on graphing procedures to each student. Individually, in small groups, or as a class read and discuss the material. Examples of graphs clipped from newspapers and magazines may be useful if your students are relatively unfamiliar with graphs. Actively engage students' minds in preparation for the process skill activity by asking the following questions.

- Can you think of a handy mnemonic for remembering which type of graph to use when?

- Why is precision important when creating a graph?

- Explain the following statement. Graphs depict not only information but also relationships.

Process Skill Activity

Students should work individually on this activity to give them maximum practice. You might want to reproduce the worksheets as transparencies to be used on an overhead projector.

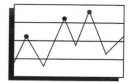

Students will use the data tables created in the last activity as the basis for making line and bar graphs.

Graphing Procedures

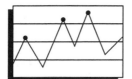

Graphs convert data sets into pictures, which are often better understood than narratives or columns of numbers. Line graphs are often used when data has taken place over time, such as population changes or one person's growth. Bar graphs are often used for descriptive data, such as car colors or food preferences of a given population.

General Procedures for Line and Bar Graphs

1. Draw a horizontal and a vertical line that meet at a right angle. The horizontal line is the x-axis and the vertical line is the y-axis.

2. Look at the data set to be graphed. Determine the manipulated variable (MV) and the responding variable (RV). The manipulated variable is written on the horizontal or x-axis. The responding variable is written on the vertical or y-axis. (Writing is more legible when written horizontally, even to label the vertical axis.)

3. Write an appropriate title for the graph at the top of the page. The title should contain the names of both variables.

4. Decide whether you will need a line graph or a bar graph. An easy way to figure this out is to look at the manipulated variable. If the manipulated variable represents passage of time, you will need a line graph. If the manipulated variable is types of things, you will need a bar graph.

Name _____   Date _____

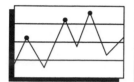

Making Line Graphs

1. Determine how to number the x-axis and y-axis. These numbers are called *increments*. While the increments for the x-axis and y-axis may be numbered differently, the numbering progression on each axis must be consistent. Looking at the range (highest and lowest numbers) for your data will help determine how to set up the increments for each axis. Make sure the graph fills the page.

2. Number the graph. It is not necessary to number every increment.

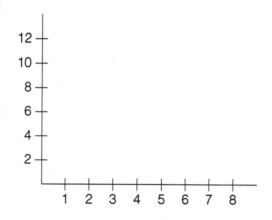 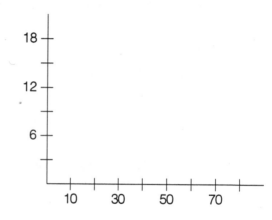

3. Plot your data. For each manipulated variable number in your data there will be a corresponding responding variable number. This matched pair of numbers represents the coordinates for one point on the graph line. Find the manipulated variable number on the x-axis. Then move up that line until you find the responding variable number on the y-axis. Place a point at the location. The point can be represented by (X,Y) or (MV, RV). This process is repeated for each pair of numbers.

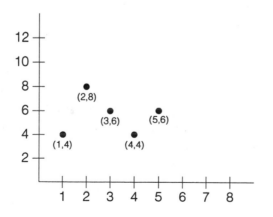

Teaching Science Process Skills © 1995 Good Apple

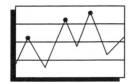

4. Connect the points on the line graph.

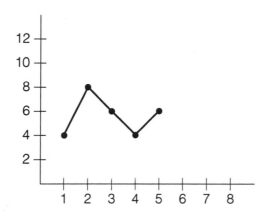

Making Bar Graphs

1. Number the increments for the responding variable on the y-axis only.

2. Write the categories for the manipulated variable on the x-axis.

3. Decide how wide each bar should be.

4. The height of each bar should correspond with the number counted on the y-axis.

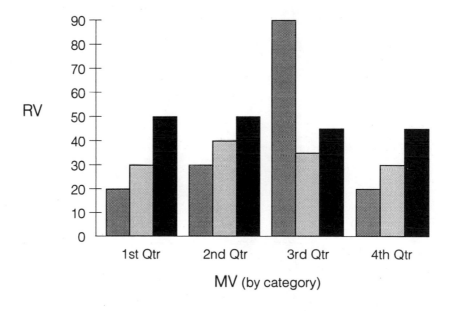

Graphing

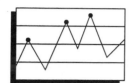

Use the six data tables you created in the previous activity to make line or bar graphs. Remember to give each graph a title and to label each axis. Fill the space as completely as possible.

Investigation 1

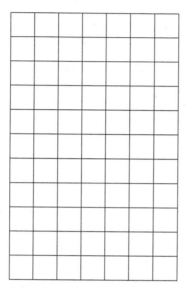

Investigation 2

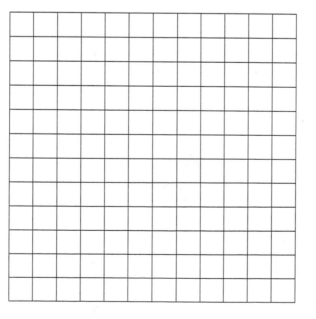

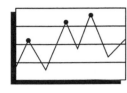

Investigation 3

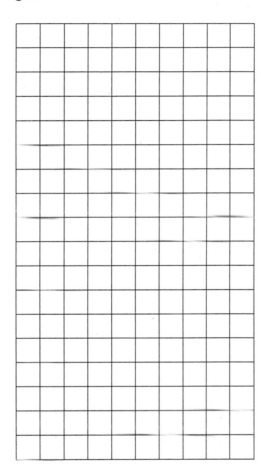

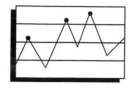

Investigation 4

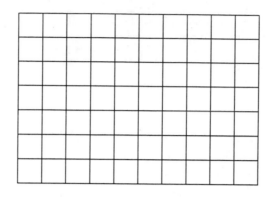

Investigation 5

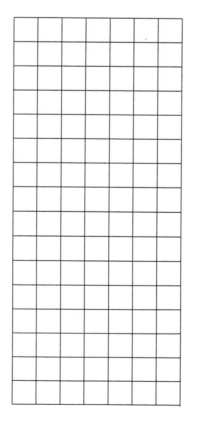

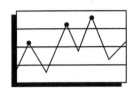

Investigation 6

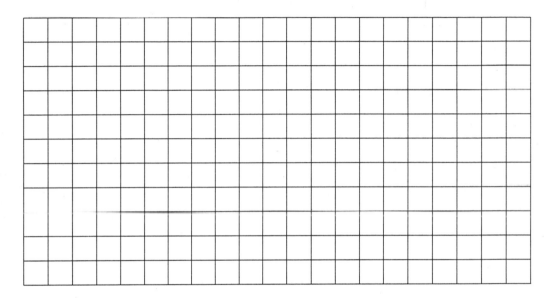

Data Tables and Graphs

Materials (per student)

- ruler
- worksheet on pages 133-134
- graph paper on pages 135-136
- additional materials needed for students to conduct their own investigations

Active Engagement Before the Task

Review the procedures for making data tables and graphs. Actively engage students' minds in preparation for the process skill activity by asking the following questions.

- What purposes can you think of for a double or triple line graph?

- What type of information would you expect to find on a double or triple bar graph?

- True or false—graphs contain qualitative as well as quantitative information. Explain your answer.

Process Skill Activity

Students should work individually on this activity for maximum practice. Supply worksheets and blank graph paper for each student. For the original investigations in Part Three, students may wish to work in pairs on data collection and tabulation. At the completion of the activity, invite students to share their findings.

In this activity, students will make data tables and use them to create single and double line graphs. In addition, students will conduct their own investigation in which they will collect data, create two data tables, and make a line and a bar graph.

Name _____ Date _____

Data Tables and Graphs

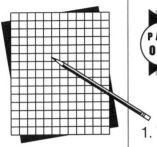

PART ONE

This activity will give you more practice making data tables and graphs.

Create a data table below for each data set. Next, determine whether a line graph or a bar graph would best represent the data in each set. Finally, create the appropriate graph on graph paper. Write a title for each graph and label the x-axis and y-axis.

1. The number of bears was counted in a park over time.

    1930 - 45 bears, 1940 - 41 bears, 1950 - 35 bears, 1960 - 30 bears, 1970 - 26 bears, and 1980 - 19 bears.

2. The height of a tree was measured over a number of years.

    4 years - 1 meter, 8 years - 2 meters, 12 years - 3 meters, 16 years - 9 meters, 18 years - 16 meters.

PART TWO The following data can be represented by a double-line graph. First create a data table with the two sets of data. The two sets of data can then be represented on one graph on a piece of graph paper.

A bank robber flees a bank running at a top speed of 20 feet per second. The getaway car is 60 feet away. The car begins to move away from the bank at the same moment the robber leaves the bank. The getaway car accelerates uniformly at a rate of 4 feet per second. Will the robber overtake and be able to enter the getaway car?

Robber Data
0 seconds - 0 feet, 1 second - 20 feet, 2 seconds - 40 feet, 3 seconds - 60 feet, 4 seconds - 80 feet, 5 seconds - 100 feet, 6 seconds - 120 feet

Car Data
0 seconds - 60 feet, 1 second - 64 feet, 2 seconds - 68 feet, 3 seconds - 72 feet, 4 seconds - 76 feet, 5 seconds - 78 feet, 6 seconds - 82 feet

Did the robber escape?_____

PART THREE You're on your own! Your task is to collect two data sets of your choice. Choose the variables, plan how to measure them, record the measurements in two data tables, and graph the data. Select your investigations so that one graph will be a line graph and the other will be a bar graph.

Here are some suggestions:
- left- and right-handed classmates by gender
- heart rate after various exercises
- body temperature at intervals during a class period
- time of sunrise and sunset for a week

Teaching Science Process Skills © 1995 Good Apple

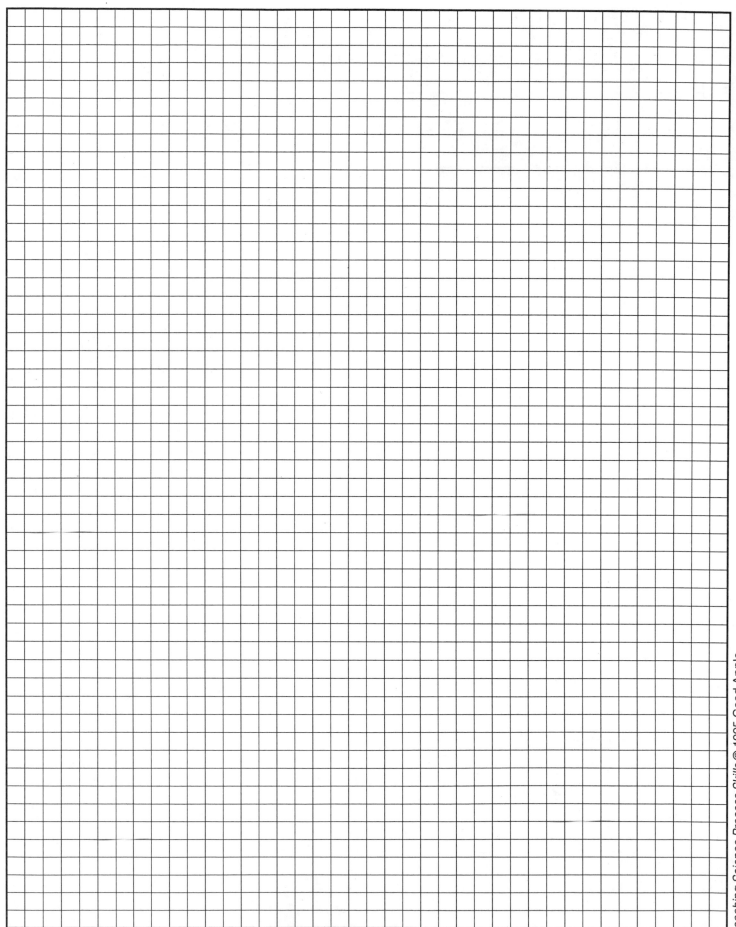

Teaching Science Process Skills © 1995 Good Apple

Using Graphs to Make Predictions

Materials (per student)

• worksheet on pages 138-139

Active Engagement Before the Task

Actively engage students' minds in preparation for the process skill activity by asking the following questions.

• How do you think a graph might be used to make a prediction?

• Can you count on such a prediction to be accurate? Why or why not?

• What types of predictions would lend themselves to being made on a graph?

Process Skill Activity

Students will work individually on the graphing portion of this task. You may wish to have them work in pairs or small groups to discuss the follow-up questions. Be sure students understand the difference between interpolated and extrapolated predictions before beginning the activity. Use examples if you feel it's appropriate. Invite students to share their responses to the questions.

In this activity, students will be introduced to interpolated and extrapolated predictions. They will create a graph, make both kinds of predictions, and respond to some questions about the process.

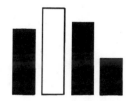

Using Graphs to Make Predictions

An interpolated prediction is one that is made between known data points. An extrapolated prediction is one that is made outside or beyond known observations.

The data table below represents the increase in stretch of a hanging spring each time a weight is added. All weights are equal. Study the data table. Note that some spaces are left blank. Create a graph for this data. Then make interpolated and extrapolated predictions and record them on the data table.

Objects of Equal Mass	Distance of Spring Stretch
0 objects	0 cm
2 objects	2 cm
3 objects	
4 objects	4 cm
5 objects	
6 objects	6 cm
7 objects	
15 objects	

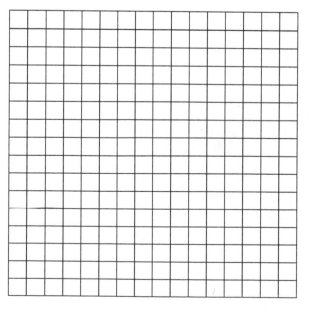

Teaching Science Process Skills © 1995 Good Apple

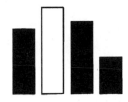

▼ Answer the following questions about predictions.

1. An interpolated prediction is one that is made between known data points. Which of the predictions made on the data table are interpolated?

_____

2. An extrapolated prediction is one that is made outside or beyond known observations. Which predictions are extrapolated?

_____

3. In which of your predictions are you the most confident? Why?

_____

_____

4. In which of your predictions are you the least confident? Why?

_____

_____

5. Explain how the graph helped you with your predictions.

_____

_____

_____

_____

*T*hinking about Combustion

T e a c h e r   N o t e s

Materials **(per group)**

- oil base clay (to hold the candle in place)
- candle
- red and blue markers or pencils
- matches (used under close adult supervision)
- second hand or stopwatch
- four glass containers (½ pint, 1 pint, 1 quart, 1 gallon) that can be easily inverted over a candle
- worksheet on pages 141-144
- graph paper on pages 135-136

Active **E**ngagement **B**efore the **T**ask

Actively engage students' minds in preparation for the process skill activity by asking the following questions.

- What is the definition of combustion?

- What three variables affect combustion?

- Based on these variables, what do you think would be the most effective way to end combustion? Why do you think so?

Process **S**kill **A**ctivity

Students will work in small groups on this activity. Distribute the worksheet and review the information on combustion. Remind students of safety procedures when using matches. Review extrapolated and interpolated predictions. Have groups come to consensus on their results before inviting them to share with the class.

In this six-part activity, students will practice the skills of hypothesizing, writing an operational definition, graphing, and making predictions. If you do not have block or flexible scheduling, you may need two days to complete the activity.

Thinking about Combustion

You will be using the combustion of a candle inside various transparent containers to practice hypothesizing and writing operational definitions. Combustion is the process of burning. Things that burn, such as coal, wood, and gasoline are said to be combustible. There are three important variables that affect combustion.

Combustion Variable 1: **oxygen**

Combustion Variable 2: **heat**

Combustion Variable 3: **fuel** (something to burn)

Knowing these variables can be very helpful when extinguishing a flame. Name some ways to extinguish a flame and tell which combustion variable is affected.

Method 1 _____

Method 2 _____

Method 3 _____

 PART ONE Stand the candle upright in a small ball of clay. Light the candle. Use caution when lighting and disposing of the match. Place the smallest jar (Jar A) over the candle. Study the flame to determine when combustion ends. Repeat this at least three times.

Describe how you are determining the end of combustion.

_____

_____

What variable are you testing? _____

_____

Write an operational definition for this variable (how you will measure this variable).

_____

_____

 PART TWO In part two you will invert three additional different-sized jars (Jars B, C, and D) over a burning candle.

Jar A = ½ pint = 2 volume units   Jar C = 1 quart = 8 volume units
Jar B = 1 pint = 4 volume units   Jar D = 1 gallon = 32 volume units

The volume units will make it easier for you to graph your data.

▼ Write a hypothesis using the "If . . . , then . . ." format that predicts the relationship between jar volume and the time it takes before combustion ends.

_____

_____

▼ Test the two smallest jars (A and B). Record your findings below. Perform the test three times.

Jar A Burning Time

Trial 1: _____ seconds   Trial 2: _____ seconds   Trial 3: _____ seconds

Jar B Burning Time

Trial 1: _____ seconds   Trial 2: _____ seconds   Trial 3: _____ seconds

▼ Calculate the average burning time for Jars A and B.

Jar A: _____ seconds   Jar B: _____ seconds

▼ Create a data table for the average burning times.
(Leave room for Jars C and D.)

PART
THREE

Graph the average burning times on a piece of graph paper. Remember to label the axes and give the graph a title.

Teaching Science Process Skills © 1995 Good Apple

PART FOUR ▶ Use your graph to extrapolate a prediction for the burning time for Jar D.

_____

Place a red dot on your graph where you think this data will fall.

▶ Test your prediction for the burning time for Jar D. Perform the test three times.

Jar D Burning Time

Trial 1: _____ seconds   Trial 2: _____ seconds   Trial 3: _____ seconds.

Calculate the average burning time and record it on your data table.

_____ seconds

Plot this information on the graph in blue.

PART FIVE ▶ Use the graph to interpolate a prediction for the burning time for Jar C.

_____

Place a red dot where this data will fall.

▶ Test your prediction for the burning time for Jar C. Perform the test three times.

Jar C Burning Time

Trial 1: _____ seconds   Trial 2: _____ seconds   Trial 3: _____ seconds.

Calculate the average burning time and record it on your data table.

_____ seconds

Plot the average burning time for Jar C in blue.

PART SIX ▶ Use the data you have collected to complete the data table and answer the questions that follow. Be sure to include units. This will help you summarize what you did with the graphs.

Method of Prediction	Jar	Predicted Burning Time	Actual Burning Time	Difference
extrapolation	D			
interpolation	C			

Name _____ Date _____

1. Was your hypothesis correct? Explain.

_____

_____

2. Which type of prediction was more accurate, interpolated or extrapolated?

_____

3. Check with several classmates. Which type of prediction was more accurate for them?

_____

_____

4. Why were predictions not made for Jars A and B?

_____

_____

5. Infer why larger jars have a longer burning time than smaller jars.

_____

_____

6. Summarize the differences between an extrapolated prediction and interpolated prediction.

_____

_____

_____

_____

_____

_____

_____

Teaching Science Process Skills © 1995 Good Apple

Data Interpretation

Materials (per group)

• worksheet on pages 146-149

Active Engagement Before the Task

Make an overhead transparency of the first data table on the work-sheet and display it for the class. Cover the conclusions and infer-ences below the table. Actively engage students' minds in prepara-tion for the process skill activity by asking the following questions.

• What is a summary? How could you summarize the information on this table?

• Can you make any inferences from this information?

• What reasons could you give for the fact that different amounts of water were absorbed each time by the same towel type?

▼ In this activity, students will evaluate several conclusions and inferences derived from given data sets.

Process Skill Activity

Students will work in pairs or small groups. Distribute the worksheet and review the information on conclusions and inferences with refer-ence to the paper towel investigation. You may wish to evaluate the first conclusion (salty tasting salts produce a red color) as a class. Groups should come to consensus on their evaluations before shar-ing results with the class. Bring the class to consensus.

Data Interpretation

Tables and graphs prepare data for interpretation in the form of conclusions and inferences. A *conclusion* is a factual summary of data. Usually more than one conclusion statement is required to summarize a data set. An *inference*, on the other hand, is a generalization that explains or interprets the data set. More than one inference is often possible for a given set of data. Study the data and the interpretation examples below. The data is from an investigation that measured the absorbency of three types of paper towels.

Amount of Water Absorbed (ml)

Towel Type	Towel Size	Trial 1	Trial 2	Trial 3	Trial 4	Average
A	225 cm^2	25	28	24	31	27
B	225 cm^2	26	27	23	22	24.5
C	225 cm^2	18	20	23	21	20.5

Conclusions Towel A absorbed an average of 27 ml of water. Towel B absorbed an average of 24.5 ml of water. Towel C absorbed an average of 20.5 ml of water.

Inferences Towel A was the most absorbent of the towels tested. Towel C was the least absorbent.

An experiment tested nine different salts. Colors of the salt were noted. The salts were tasted and heated in a flame. The color of the flame produced by each salt was recorded.

Salt	Color	Taste	Flame Color
A	blue	salty	red
B	blue	biting	yellow
C	blue	biting	red
D	green	biting	red
E	green	none	green
F	green	biting	red
G	white	biting	red
H	white	none	yellow
I	white	salty	red

Teaching Science Process Skills © 1995 Good Apple

Evaluate the validity of the following conclusions and inferences.

Conclusion: Salty tasting salts produce a red color.

Evaluation _____

_____

_____

Conclusion: No two blue salts are exactly the same.

Evaluation _____

_____

_____

Inference: The color of salt determines the color produced when the salt is heated.

Evaluation _____

_____

_____

Inference: There is no single factor in determining if a salt will produce a yellow color when heated.

Evaluation _____

_____

_____

PART TWO To determine one of the requirements for the germination (sprouting) of one variety of seeds, Sally divided 900 seeds into groups of 100 each. Each group of seeds was placed in a different germinator under the same conditions except for temperature. The number of sprouted seeds was counted after 30 days.

Temperature °C	6°	8°	11°	13°	18°	25°	30°	35°	39°
Sprouts	0	0	0	0	16	50	84	30	10

Graph the data, so that you will be better able to interpret it.

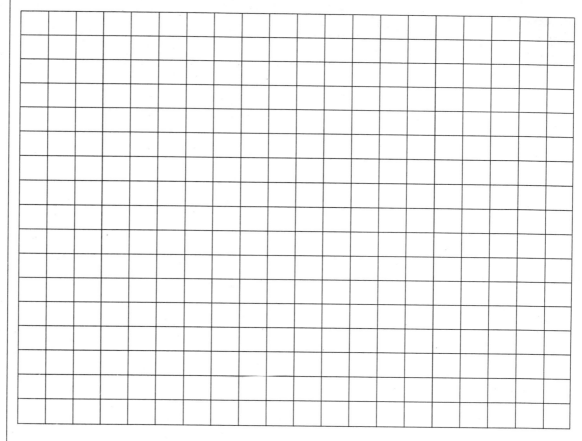

Several conclusions and inferences concerning the data have been provided. Some are not appropriate. Evaluate each conclusion and interpretation. Write explanations that defend your decision concerning the interpretations.

Conclusion: The higher the temperature the greater the number of seeds that sprout.

Evaluation _____

_____

_____

Teaching Science Process Skills © 1995 Good Apple

Conclusion: Some seeds sprout if kept between 18-39 degrees Celsius. Most will sprout at about 30°C.

Evaluation _____

_____

_____

Inference: In order for seeds to sprout, they need favorable conditions.

Evaluation _____

_____

_____

Inference: Approximately half of any kind of seeds sprout at 25°C.

Evaluation _____

_____

_____

Inference: A temperature of 13°C or below is required to stop seeds from sprouting.

Evaluation _____

_____

_____

The Kaibab Deer Story

Materials (per student)

• worksheet on pages 151-152

Active Engagement Before the Task

Actively engage students' minds in preparation for the process skill activity by asking the following questions.

• What does the phrase "balance of nature" mean? (stability of plant or animal population size within a community or ecosystem)

• What factors might upset the balance? (earthquakes, fires, floods, logging, livestock grazing, development)

• Is man or nature more responsible for destroying nature's balance? Why?

Process Skill Activity

Students will work in groups for this activity. Groups should come to consensus on conclusions and inferences. Invite groups to share their recommendations when all work has been completed.

▼ *In this activity, students will do a complete interpretation of data (collected over years) about the size of the mule deer population on the Kaibab Plateau in Arizona. In addition to graphing, making inferences, and drawing conclusions, students will make recommendations with regard to wildlife management.*

Name _____ Date _____

The Kaibab Deer Story

PART ONE ▶ You will be interpreting some historical data collected over the years about the size of the mule deer population on the Kaibab Plateau in Arizona. You will do a complete interpretation, including making graphs, drawing conclusions, and making inferences. In addition, you will make recommendations or record a plan of action suggested by your conclusions and inferences. Make a data table and graph for the following information.

Kaibab Plateau Deer Population, 1907-1938

1907 - 6,000 deer, 1910 - 10,000 deer, 1915 - 25,000 deer,
1920 - 56,000 deer, 1923 - 100,000 deer, 1925 - 85,000 deer,
1930 - 30,000 deer, 1935 - 14,000 deer, 1938 - 9,000 deer

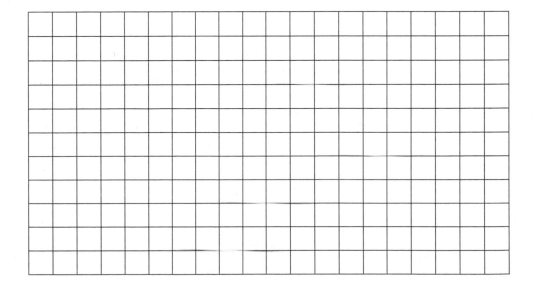

Name _____ Date _____

 PART TWO

Use the data to write the best possible conclusion. Make some inferences about your conclusion.

Conclusion _____

_____

_____

Inferences _____

_____

_____

_____

PART THREE

You are probably curious about what happened to the Kaibab deer. Between 1907 and 1923, 300 coyotes and 600 mountain lions were killed by hunters on the Kaibab Plateau. Based on this additional information, what possible inferences from the data available for the years 1907 to 1923 can be made?

_____

_____

_____

PART FOUR

During the years of increasing and decreasing deer populations, the range of the deer was so badly overgrazed that thousands of deer died from starvation. What logical recommendation or recommendations would you make based on the available information? Remember that a recommendation is an action to be taken based on findings.

_____

_____

_____

_____

Teaching Science Process Skills © 1995 Good Apple

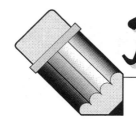

Investigating: Experiments and Surveys

O b j e c t i v e s

Based on the activities and discussion in this chapter, students will be able to:

▼ Write a detailed experiment report.

▼ Define a scientific investigation as either a survey or an experiment.

▼ Compare and contrast the differences between a survey and an experiment.

▼ Distinguish surveys from experiments in a list of investigations.

▼ Design and/or conduct an experiment that will permit an appropriate evaluation of a given hypothesis.

▼ Design and/or conduct a survey that will permit an appropriate evaluation of a given research question.

▼ Identify the important components (variables, hypothesis, operational definitions, data collection design) of a given experimental design and explain the purpose of each.

▼ Identify the important components (variables, research questions, population, sampling procedure, data collection design) of a given survey design and explain the purpose of each.

▼ Explain why "control" is an important factor in the accuracy of all investigations.

Investigating Through Experimentation

Scientists attempt to understand and explain the natural world through the empirical method of using physical evidence collected through thoughtful observation. Since natural events are complex and involve many factors or variables, scientists break down events into variables in order to study the way one variable affects another. If enough variables are studied in the investigation, clear understanding may be the result.

One way to conduct an investigation is through experimentation. In an experiment, scientists ask a question about how one variable will affect another variable. A research question defining the overall investigation is written. Often a prediction or hypothesis is made and a method is devised to test the hypothesis. The hypothesis predicts how one variable will affect another variable. A method is devised to test the hypothesis. Operational definitions for these variables are written. The test is carried out and thoughtful observations and data are recorded and organized in tables and graphs. Conclusions, inferences, and recommendations are made. The findings often lead to new questions, which lead to new and related experiments.

Teaching Science Process Skills © 1995 Good Apple

M&M Color Experiment

Materials (per group)

- M&M's of various colors
- coffee filters
- container with smaller circumference than coffee filters
- water
- scissors
- transparency of graph paper on page 135
- introductory information on investigating through experimentation on page 154
- worksheet on page 156
- experiment report model on page 173
- experiment report on page 174-175

Active Engagement Before the Task

Distribute the introductory information on investigating through experimentation to each student. Individually, in small groups, or as a class read and discuss the material. Actively engage students' minds in preparation for the experiment by asking the following questions.

- What is the definition of a *type one* research question?

- M&M's now come in three varieties—plain, peanut, and peanut butter. Which would you predict is the most popular? How could you find out?

- Can you think of an experiment related to M&M colors?

Process Skill Activity

Distribute worksheets, experiment reports, and experiment report models. Point out that the M&M experiment has a *type one* research question. Review the information on dyes and capillary action and invite the groups to set up their experiments. Note that yellow dye 5 and 6 will appear as one yellow. The numbers are used for technical accuracy. If time permits, you may want students to record observations at certain preset times. Review the format for reporting the results of their experiment. Once groups reach consensus on the question, encourage them to share their findings.

In this activity, students will perform an experiment to separate the dyes used to color M&M's. The experiment is most effective if it is set up and allowed to dissolve for 24 hours. The experiment can be done after an hour, however, if you are short on time.

M&M Color Experiment

Research Question

Which M&M's contain yellow dyes 5 and 6?

Information

Dyes are used in many food products. M&M's feature several bright colors that may contain dyes. Dyes can be separated from each other using a chromatography process. In this method, solids of different mass can be separated because lighter solids move greater distances when moistened than do heavier solids. This movement in water is called capillary action.

Procedure

Cut a tab into a coffee filter. Place the flattened filter on top of a container filled ¾ full of water. Extend the tab into the water. Gently place an M&M on top of the coffee filter. Do this for each color of candy.

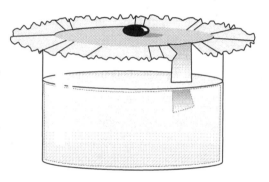

The capillary action will take at least an hour. It is more effective in 24 hours. When the water reaches the M&M, the outer coating will be dissolved and will spread out on the filter paper. The heavier dyes will remain near the center of the filter paper. The lighter ones will be carried farther away. Place a transparency of centimeter graph paper over each piece of filter paper and count the number of squares that contain yellow dye 5 and 6.

Record your results on the experiment report following the model.

Pendulum Experiments

Materials (per group)

- string
- ring stand
- washers (or any object that can be used to vary mass)
- meter stick
- second hand
- worksheet on page 158
- experiment report model on page 173
- experiment report on page 1/4-175

Active Engagement Before the Task

Actively engage students' minds in preparation for the experiment by asking the following questions.

- Foucault's pendulum furnished proof of the earth's rotation for the first time. What do you know about Edgar Allen Poe's pendulum?

- How do you think a pendulum makes a clock work?

- Galileo timed a swinging chandelier with his pulse while lying on his back in a cathedral in Pisa, Italy. Can you think of some other ways to time a pendulum without using a clock or watch?

Process Skill Activity

Students will work in small groups on this activity. Distribute the worksheet, experiment report and experiment model. Review the assignment and emphasize the importance of controlling the variables not being tested. Review the report format if necessary. Encourage groups to put together a formal presentation of their experiments and results. These presentations should include a visual and should represent the work of the entire group.

This experiment contains a type two research question. In this activity, students will design a series of experiments to test how three variables will affect the swing of a pendulum.

Pendulum Experiments

A pendulum is an object connected to a fixed point by a string, wire, or rope. When set in motion, a pendulum swings back and forth. You have heard about Galileo and Foucault, whose pendulum experiments made important contributions to science. Visualize a pendulum. What variables might affect its swing? List them below.

Variable 1 _____

Variable 2 _____

Variable 3 _____

Design a series of experiments that will test how each of the variables listed above might affect a pendulum's swing or frequency. The frequency of a pendulum is the time it takes to make a complete cycle (from starting point back to starting point). You will need at least three experiments—one for each variable. This will require you to write a *type two* research question and a hypothesis for each experiment. Controlling all variables not being tested is extremely important in this series of experiments.

Teaching Science Process Skills © 1995 Good Apple

Your Own M&M Survey

Materials (per group)

- package of plain M&M's
- introductory information on investigating through surveys on pages 160-162
- worksheet on pages 163-164

Active Engagement Before the Activity

Distribute the introductory information on investigating through surveys. Individually, in small groups, or as a class read and discuss the material. Actively engage students' minds in preparation for the process skill activity by asking the following questions.

- Which color of M&M do you predict you'd find most often in a full bag? Why do you think so?

- How do you think the manufacturers of M&M's decided on what colors to use? Do you agree with their choices?

- If you could add a color to the M&M spectrum, which one would you add and why?

Process Skill Activity

Students will work in pairs or small groups. Remind them to refer to the survey information as they complete their work. Have groups share their results. Once all results are tabulated, see what further conclusions can be drawn.

In this activity, students will conduct a survey to find out how often various colors occur in a bag of M&M's. They will graph the results, draw some conclusions, and make inferences about their conclusions.

Investigating Through Surveys

Not all science research is done in a lab. Not all scientific investigations are done through experimentation. A great deal is done in the field and in the community. Because some variables cannot be manipulated in the lab, surveys are an extremely important investigative method.

A survey is a critical inspection used to provide exact information. It is a way of discovering information in or from a particular area. Information in a survey is observed directly. For example, if you wanted to survey the number and location of leaky faucets in your school, you would move from one faucet to another as you made observations. You would not ask the custodian. Surveys use *type one* research questions.

Consider the following examples of surveys:

- A team of students surveyed the litter in four sections of their city.

- A student made basic physical measurements of a beach in his town.

- Hawks were counted during their winter migration in Harris County.

- Students counted the unoccupied school rooms that had lights left on.

- A team of specialists counted the number of dinosaur bones found in a square mile in Montana.

The following example illustrates several of the components of a survey. The students who conducted the survey decided to directly observe M&M's in a way that would provide the information needed to answer their questions.

Teaching Science Process Skills © 1995 Good Apple

Middle School Survey

Some middle school students were at lunch in the cafeteria. Their talk turned to the M&M's one of the students was eating. One student said that he was surprised at the large number of red M&M's in the bag. Another said that she had never observed red M&M's. The conversation led to a number of questions.

- How many M&M's are found in a typical bag?

- How many colors of M&M's are there?

- Why were those particular colors used?

To answer some of these questions, the students decided to do an investigation using a survey. They each agreed to buy two bags of plain M&M's at a store near their homes and count the contents the next day at lunch.

Survey Components	Example from M&M Survey
Variables	Several variables were identified • number of pieces per bag • number of colors • number of each color
Research Questions	*Type one* research questions were asked. The questions focused on the variables that had been identified. Sometimes predictions were made.
Population (variable to be observed)	Plain M&M's, not peanut or peanut butter candies.
Area (region in which population will be observed)	Students' neighborhood or school boundary.
Sampling Method (method for observing population)	The students probably would not have the time or resources to observe the entire population—every bag of plain M&M's—within their area. Therefore, it is important that the sample is an accurate representation of that population.

Teaching Science Process Skills © 1995 Good Apple

A sample must be selected in a way that insures representation of the populations. The sampling process, therefore, is extremely important. There are three basic sampling strategies that have major differences in their ability to fairly and accurately represent a population. The chart below defines these differences.

Sample Type	Examples from M&M Survey
Random (Excellent system of selection. Every member of the population has an equal chance of being selected)	Write each store selling M&M's in the given area on a slip of paper and randomly draw 20% from a jar.
Systematic (Acceptable system of selection. Collects observations by a pre-determined system or pattern. Not every member of the population has an equal chance of being selected.)	Select every fifth store selling M&M's in a given area from a list until you have 20% of the total.
Convenience (Weak system of selection. Consists of data that is easy to collect. Cannot be assumed that samples of convenience are representative of the population.)	Identify stores that are near school or home.

Name _____ Date _____

Your Own M&M Survey

Conduct your own M&M survey in response to the research question below.

Research Question: To what extent do various colors occur in a bag of M&M's?

Sampling Method

_____

_____

_____

Procedure for Collecting Data

_____

_____

_____

Qualitative Observations

_____

_____

_____

Data Table

Teaching Science Process Skills © 1995 Good Apple

Name _____ Date _____

Graph

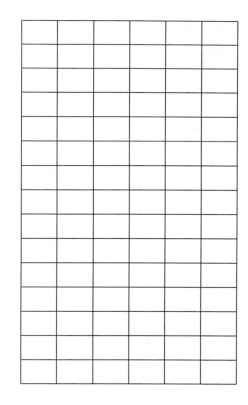

Conclusion(s) about M&M's

_____

_____

_____

_____

Inferences

_____

_____

_____

_____

_____

Teaching Science Process Skills © 1995 Good Apple

Cereal Survey

Teacher Notes

Materials (per group)

- worksheet on pages 166-167
- various cereals or cereal data
- notebook paper

Active Engagement Before the Task

Review survey and sampling techniques. Explain the cereal survey to students. Actively engage students' minds in preparation for the process skill activity by inviting them to create a system that will insure that no one in their group reports on the same cereal as someone else in their group. Allow ten or fifteen minutes for this activity and have students report their decisions to the class. Set a reasonable deadline for acquisition of cereal data.

Process Skill Activity

Students will work in small groups. Distribute the worksheet and encourage groups to come to consensus on the fourth research question required on the sheet. There should also be consensus concerning all survey procedures. Once the results of the survey are in, bring the class to consensus on the three research questions done by each group. Individual groups may also share the results of their fourth question.

▼ *This activity requires students to go to a grocery store to survey cereal content. If each member of a small group covers four or five different cereals, the group should have enough data to complete a meaningful survey. Survey results will be tabulated and reported to the class.*

Name _____ Date _____

Cereal Survey

You are familiar with breakfast cereals. But what do you really know about the nutritional quality of these products? This activity will give you the opportunity to conduct a survey on breakfast cereals. Here are some research questions that may assist you in conducting your survey.

1. Which breakfast cereal(s) contains the least amount of sodium per ounce?

2. Which breakfast cereal(s) contains the least amount of sugar per ounce?

3. Which breakfast cereal(s) contains the most dietary fiber per ounce?

4. Write a fourth research question.

_____

_____

PART ONE ► Use the following procedures to guide you as you conduct your survey.

What is your research question? (What will be the focus of your survey?)

_____

_____

What is the population of the survey? (What cereals will be the target of the survey?)

_____

_____

_____

What is the area of the survey? (Where will the survey be conducted?)

_____

_____

_____

Teaching Science Process Skills © 1995 Good Apple

How will the population be sampled? (How will the sample be obtained?)

_____

_____

_____

_____

What are the variables and other information needed to answer the research question? (Exactly what information will be collected during the survey?)

_____

_____

_____

_____

How will the data collected be organized?

_____

_____

_____

_____

What other decisions will you need to make?

_____

_____

_____

_____

 Once you have organized your data, conclusions, inferences, and recommendations, prepare a presentation for the class.

Your Own Great Investigation

Teacher Notes

Materials (per student)

- worksheet on pages 169-170
- guidelines for independent research on page 171
- self-evaluation of independent investigation on page 179
- experiment plan and approval on page 172 and experiment report on pages 174-175
  or
- survey plan and approval on page 176 and survey report on pages 177-178

The Great Investigation

Distribute the guidelines for independent research. Individually, in small groups, or as a class read and discuss the material thoroughly. Students may work alone or in pairs on the investigation. Stipulate reasonable deadlines and post them clearly. Once students have decided on a topic for investigation, have them complete the experiment or survey plan and approval for your review. Assist them in brainstorming and acquiring all materials needed. Encourage students to conduct a library search. Students should complete the self-evaluation of independent investigation when they finish their investigations.

▼ *In this activity, students will have the opportunity to apply the process skills they have learned in their own investigations. Students will be required to choose their own topics and design their own experiments. In addition, each student will keep a journal and submit a final report using the prescribed format.*

Name _____ Date _____

Your Own Great Investigation

Here is an opportunity to apply all the science process skills you have learned. You will ask an original question and conduct an experiment or survey to answer that question.

Your Task

1. Ask a question about a physical or biological event or relationship. Make sure the topic offers the potential for experimentation. (Flying saucers won't do.) Consumer testing is a good area for first-time projects. The best topics can arise from your hobbies, interests, activities, and skills.

2. Decide whether you will perform an experiment or a survey.

3. Complete the experiment plan and approval or survey plan and approval. Meet with your teacher to discuss your plan and get final approval.

4. List the potential variables. Choose a manipulated variable and a responding variable. Operationally define the variables so they can be measured.

5. Write a specific research question.

6. Write hypotheses that provide an exact focus for the experiment.

7. Conduct a library search and write a review of the literature about your topic. A literature review summarizes information about your topic into a report.

8. Design an experiment to collect data that will answer the research question and hypotheses. Remember to control all variables except the manipulated and responding variables.

9. Write the procedure for your experiment. You will probably want to build in more than one test of variables. You may want to repeat the same experiment several times. With repeat testing, you will have more confidence in your findings.

10. Gather materials.

11. Do the experiment. Remember, more data is better than less data.

12. Compile the results. Quantitative data should be recorded in data tables and you will want to include graphs to help with interpretation. Don't forget to record qualitative data also.

13. Interpret the data. Write conclusions, inferences, and discussion. Make recommendations. Don't be too general. For example, if you test only one type of radish seed, your interpretation must refer to only that variety of radish seed.

14. Prepare a final report. Use the experiment or survey report. You will be held accountable for all critical information and techniques.

Common Problems in Student Experiments

- Low sample size.

- Lack of control of everything but the manipulated and responding variables.

- Inaccurate measurement.

- Experimenter bias. Keep an open mind.

- Population selection bias.

Suggestions For Your Topic

- How many seeds are found in common types of fruits?

- How does the shape of a container affect the evaporation rate of liquids?

- Which pill design—tablet, caplet, or capsule—dissolves faster?

- Which bottle designs are the most childproof or tamperproof?

- Which heats liquids faster—gas, electric, or microwave heat sources?

- How do mealworms respond to selected stimuli, such as temperature, odor, or light?

- How does color affect the surface temperature of construction paper?

- What percent of a dissolving antacid tablet is a gas?

Teaching Science Process Skills © 1995 Good Apple

Guidelines for Independent Investigation

Your final product must be typed or clearly written in ink.

Clearly describe manipulated and responding variables.	20	points
Clearly state your research question and hypothesis. You must be able to answer the question with your research.	20	points
Clearly and completely state your procedure.	20	points
Identify all controlled variables.	15	points
Include clearly written observations, data tables, and graphs.	30	points
Write clear conclusions.	30	points
Write specific and clear inferences about conclusions and discuss your research clearly.	30	points
Write clear recommendations.	30	points
Total Possible:	195	points

Name _____ Date _____

$\mathcal{E}$xperiment Plan and Approval

Investigation Topic _____

Manipulated Variable _____

Responding Variable _____

Research Question

_____

_____

Hypothesis/Hypotheses

_____

_____

_____

_____

Data Collection Plan

_____

_____

_____

_____

How many trials will be used in your experiment? _____

How will you control all variables except the manipulated and responding variables?

_____

_____

What special equipment will you need? _____

_____

Teacher's Approval_____

Teaching Science Process Skills © 1995 Good Apple

Experiment Report Model

Title

Write the name of your investigation.

Research Question

Communicate what variables or relationship between variables will be investigated.

Manipulated Variable

Write the manipulated variable and its operational definition.

Responding Variable

Write the responding variable and its operational definition.

Hypothesis or Hypotheses

Communicate your predictions about how the manipulated variable will affect the responding variable.

Plan or Procedure

Communicate a step-by-step process for completing the experiment. Include materials, time tables, and any other information that is important to the investigation. The plan should be detailed enough for another investigator to duplicate the experiment.

Controlled Variables(s)

Identify all factors that must be controlled to limit sources of error.

Observations, Data Tables, Graphs

Include qualitative observations.

Conclusion(s)

Write factual summaries about what happened.

Inference(s) about Conclusion(s)

Interpret, explain, and discuss the relationship between the variables. What knowledge was gained and what does it mean?

Recommendation(s)

What subsequent actions could or should be taken. Also, suggest methods for improving the experimental techniques.

Name _____ Date _____

$\mathcal{E}$xperiment Report

Title _____

Research Question

_____

_____

Manipulated Variable _____

_____

Responding Variable _____

_____

Hypothesis or Hypotheses

_____

_____

_____

_____

Plan or Procedures

_____

_____

_____

_____

_____

Controlled Variables(s) _____

_____

_____

_____

_____

Teaching Science Process Skills © 1995 Good Apple

Name ______________________ Date ______________

____

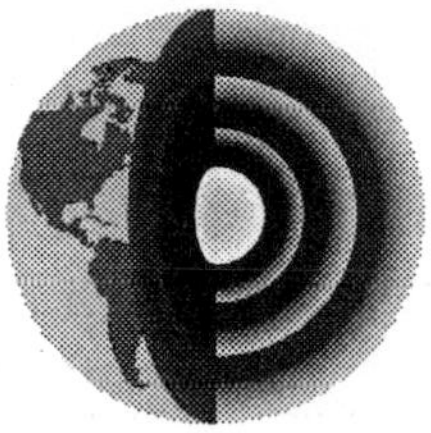

Observations, Data Tables, Graphs

____

____

____

____

Conclusion(s)

____

____

____

____

Inference(s) about Conclusion(s)

____

____

____

____

Recommendation(s)

____

____

____

____

____

Name _____ Date _____

Survey Plan and Approval

Investigation Topic _____

Population to be Surveyed _____

Area to be Surveyed _____

Sampling Method

_____

_____

Variables

_____

_____

Research Question

_____

_____

Hypothesis/Hypotheses

_____

_____

Data Collection Plan

_____

_____

How will you control variables?

_____

_____

_____

Teacher's Approval _____

Teaching Science Process Skills © 1995 Good Apple

$\mathcal{S}$ urvey Report

Name of Survey_____

Research Question

_____

_____

Variables_____

Population to be Surveyed

_____

_____

Area to be Surveyed

_____

_____

Sampling Method

_____

_____

_____

_____

_____

Name _____   Date _____

_____

Hypothesis

_____

_____

_____

_____

_____

_____

Plan or Procedures

_____

_____

_____

_____

_____

_____

_____

_____

_____

Teaching Science Process Skills © 1995 Good Apple

Self-Evaluation of Independent Investigation

You have spent a number of days performing your independent investigation. Please complete the following self-evaluation.

1. Were you working alone or with a partner?_____

If with a partner, who was your partner? _____

2. List at least five things you thought about as you worked on this project.

_____

_____

_____

_____

_____

3. List five things you did to complete this project.

_____

_____

_____

_____

_____

4. What did you like best about this project?

_____

5. What was the hardest part of the project?

_____

6. What did you learn? Explain your new understandings.

_____

_____

_____

7. What would you change about this project?

_____

_____

Assessment

Assessment is an important part of any science program.

▼ Assessment provides information about the quality and quantity of content, skills, and behaviors that a student or a group of students has attained.

▼ Self-assessment is essential if improvement is desired.

▼ Assessment that does not lead to learning is wasted time and effort.

▼ Students can use assessment instruments.

▼ Content and reasoning must be given at least equal emphasis in assessment.

▼ Teachers and students must reflect on what behaviors are appropriate.

Use the test on pages 182-188 for pre- and post-testing.

Name _____ Date _____

$\mathcal{S}$cience Process Skills Test

Circle the letter next to the most appropriate answer.

1. Which of the following is an observation only?

   A. The piece of metal is red, so it must be hot.

   B. The street is wet, so it must have rained.

   C. The table looks like it is made of wood.

   D. The child's block is orange.

2. Which of the following could be observed with the sense of sight?

   A. The temperature of the air.

   B. The change in height of plants.

   C. The sweetness of a new chemical.

   D. The sound made by an engine.

3. Which of the following is not a characteristic of an orange?

   A. It is round.

   B. It is juicy.

   C. It is red.

   D. It can be peeled.

4. A scientist discovered a large boulder in a field. The boulder was different than any of the surrounding rock in the landscape. This discovery could be labeled:

   A. a hypothesis

   B. an inference

   C. a discrepancy

   D. an interpretation of data

Teaching Science Process Skills © 1995 Good Apple

Name _____ Date _____

5. Here is some information about students in Maple School.

Name	Gender	Birthday	Nationality	Year entered school
M. Formichelli	Female	June 1974	Italian	1986
B. Thermal	Male	March 1974	Indian	1986
A. Siddiqui	Male	December 1973	Pakistani	1986
R. Johnson	Female	May 1974	Swedish	1986
R. Ali	Male	October 1973	Indonesian	1986
J. Martinez	Male	August 1973	Hispanic	1986

Which of the following categories would not allow you to separate these students into at least two different groups?

    A. gender (male or female)

    B. year of birth

    C. nationality

    D. year entered school

6. Which one of the statements below accurately characterizes an observation?

    A. serves as an explanation for an inference

    B. must involve the use of at least one of the human senses

    C. categorizes objects and/or events according to similarities

    D. serves to explain data taken in by any or all of the human senses

7. For an observation to take place, the observer:

    A. simply has to experience an object or event

    B. has to record the observed data

    C. must think about what is being observed

    D. must make an explanation of the object or event

8. A written statement of a hypothesis must contain or strongly imply which of the following variables?

A. only the independent or responding variable

B. only the dependent or manipulated variable

C. both the manipulated and responding variables

D. both the manipulated and responding variables, as well as all the controlled variables

9. The following four statements concern hypotheses.

1. A hypothesis may be formed on the basis of observation alone.

2. A hypothesis is stated in testable terms.

3. A hypothesis may be formed from inference(s).

4. A hypothesis can establish the basis for experimentation.

Which of the four statements are true about hypotheses?

A. 1, 2, 3 and 4

B. only 2, 3 and 4

C. only 2 and 3

D. only 1, 2, and 3

10. A student wants to know the effect of acid rain upon a fish population. She takes two jars and fills each of the jars with the same amount of water. She adds fifty drops of vinegar (acid) to one jar and adds nothing extra to the other. She then puts 10 similar fish in each jar. Both groups of fish are cared for (oxygen, food, etc.) in identical fashion. After observing the behavior of the fish for a week, she makes her conclusions. What would you suggest to improve this experiment?

A. Prepare more jars with different amounts of vinegar.

B. Add more fish to the two jars already used.

C. Add more jars with different kinds of fish and different amounts of vinegar in each jar.

D. Add more vinegar to the two jars already in use.

Teaching Science Process Skills © 1995 Good Apple

11. Recently, Beth heard sirens roaring on a nearby street. The next day when she went to school she saw a house covered with wide black spots and smoke. The most reasonable inference that she could make when describing what she saw was:

A. The house was destroyed by a tornado.

B. The house was destroyed by a wild animal.

C. The house was destroyed by a fire.

D. The house was destroyed by a hurricane.

12. During the night, Steve was awakened by a thunderstorm. Walking to school the next day, he saw a large tree blocking the street. The best inference that he could make is that the tree was:

A. hit by a bulldozer

B. bombed by an airplane

C. knocked down by the storm

D. destroyed by a fire

13. Which month do you think will be the coldest?

A. June

B. September

C. April

D. January

14. Dan and Dawn want to know if there is any difference between the mileage expected from bicycle tires from two different manufacturers. Dan will put one brand on his bicycle and Dawn will put the other brand on her bicycle. Which of the following variables would be the most important to control in this experiment?

A. the time of day the test is made

B. the number of miles traveled by each type of tire

C. the physical condition of the bike rider

D. the weather condition

E. the weight of the bicycle used

Name _____ Date _____

15. A group of students conducted an experiment to determine the effect of heat on the germination (sprouting) of sunflower seeds. Which of the variables listed below is the least important to control in this experiment?

A. the temperature to which the seeds are heated

B. the length of time the seeds are heated

C. the type of soil used

D. the amount of moisture in the soil

16. Highest daily temperature recorded each day for a week is shown on the data table.

Sunday	Monday	Tuesday	Wednesday	Thursday	Friday	Saturday
8°C	7°C	0°C	15°C	23°C	21°C	19°C

Which of the following statements is correct?

A. Monday had the lowest temperature.

B. The highest temperature was recorded on Thursday.

C. It snowed all day on Friday.

D. The temperature was higher on Wednesday than on Saturday.

E. Saturday had the highest temperature of the week.

17. Bob set up two identical bowls. Both contained sugar water and were open to the air. One was placed in the dark, while the other was placed in the light. What is the one item that is different from one set-up to the other?

A. the exposure to light

B. the shape of the bowl

C. the exposure to air

D. the amount of sugar in each

Teaching Science Process Skills © 1995 Good Apple

18. Which of these statements best represents a hypothesis?

A. The magnet picked up twelve paper clips.

B. The milk in this bottle froze in twenty minutes.

C. Most liquids expand when heated because the particles that make them up move farther apart.

D. The leaves on that maple tree have all turned red.

E. If the rate of the water flow continues, the pool will fill in ten minutes.

19. Which of the following Is written as an operational definition?

A. The pool froze because the temperature went below 0°C.

B. The temperature will be determined by using a Celsius thermometer.

C. How long will it take for the pool to freeze?

D. If the air temperature drops below freezing, then the pool will freeze.

20. If the amount of carrots fed to hamsters is increased, fewer will die of tail rot. The above statement is best described as:

A. an observation

B. a discrepancy

C. an inference

D. a hypothesis

21. What unit would best measure the distance between Fort Worth and Houston?

A. meter

B. kilogram

C. centimeter

D. kilometer

Teaching Science Process Skills © 1995 Good Apple

22. Students created a data table showing the kinds of candy in a grocery store. They wanted to make a graph, so they could show the data in a second way. What kind of graph is most appropriate for the data?

A. line graph

B. bar graph

C. both kinds of graphs

D. another kind of graph

23. Your lab group has made a number of observations. Your next step might be to:

A. Turn in the lab report.

B. Generate a graph.

C. Draw a picture illustrating an observation.

D. Make some inferences about your observations and investigate.

24. When converting from one unit to another in the metric system:

A. A series of prefixes are used to indicate the new unit.

B. A series of suffixes are used to indicate the new unit.

C. Conversion formulas must be used.

D. There are twelve centimeters in each meter.

25. If you were measuring volume of water, which unit would be appropriate?

A. meter

B. millimeter

C. milligram

D. milliliter

Teaching Science Process Skills © 1995 Good Apple

Answer Key for Science Process Skills Test

1. D	6. B	11. C	16. B	21. D
2. B	7. C	12. C	17. A	22. B
3. C	8. C	13. D	18. E	23. D
4. C	9. A	14. B	19. B	24. A
5. D	10. A	15. C	20. D	25. D

Portfolios and Journals

Portfolio assessment is the examination of a collection of student work that focuses on a student product and on student growth over time. Journals can be helpful for students to record their thoughts and reflect on what they have learned. The following pages provide some forms you may find helpful as you collect and assess student work.

Name _____ Date _____

$\mathcal{A}$ssignment Sheet

Assignments	Grades	Comments About Assignments
Week 1		
Week 2		
Week 3		
Week 4		
Week 5		
Week 6		

Teaching Science Process Skills © 1995 Good Apple

Journal

What skills have you mastered? What skills do you still need to perfect?	What comments do you wish to make about the week? Were the activities of value to you? What would you change? What problems did you have?
Week 1	
Week 2	
Week 3	
Week 4	
Week 5	
Week 6	

Journal

What did you learn? Explain your ideas in detail. "I learned that stars are different colors" is inadequate. Explain the how and why of what you learned. Your writing should reflect and convey thorough understanding.

Week 1

Week 2

Week 3

Week 4

Week 5

Week 6

Teaching Science Process Skills © 1995 Good Apple